Jean-Marc Chapallaz, Jacques Dos Ghali,
Peter Eichenberger, Gerhard Fischer

Manual on Induction Motors Used as Generators

Springer Fachmedien Wiesbaden GmbH

vieweg

The authors:

Jean-Marc Chapallaz is a Swiss electro-mechanical engineer specialized in the design of hydrau machinery and equipment. He worked in industry both at home and abroad for over ten years. In 1981, established his own company of consulting engineers, mainly engaged in energy and water sup projects. He has designed and implemented around 20 micro-hydropower plants ranging from 0,5 to 450 power output.

His engagement in development cooperation includes short term consultancy missions for hydropov projects in Asia and the Caribbean, the organization of training courses for hydropower engineers a other contributions to programmes of technology transfer.

Jacques Dos Ghali, is a senior lecturer at the Swiss Federal Institute of Technology in Lausanne. After years in industry (electrical power stations and networks) he was appointed to the electricity departmen the Federal Institute of Technology in 1970. His activities in the field of micro-hydropower include development of an electronic load controller and lectures on electrical engineering for hydropower train courses. His involvement in development cooperation is mainly in the field of energy planning, establishment of master plans and education programmes, and the implementation of electric laborator in universities and technical colleges in various countries in Africa.

Peter Eichenberger has been working with J.-M. Chapallaz, Consulting Engineers since 1990. He ha background in civil engineering and hydraulics, and spent a number of years with other Swiss c engineering consultants mainly on rural development projects. Prior to joining J.-M. Chapallaz, he work as a senior development adviser in a pump lift irrigation project in Sumatra, Indonesia.

Gerhard Fischer is a German mechanical engineer specialized in hydraulic machinery and govern systems. After graduating from the Institute of Hydraulic Machinery in Stuttgart, he spent six years a researcher there, working on improved turbine and governor designs for both industrialized a developing countries. In 1988, he joined J.-M. Chapallaz, Consulting Engineers where he cooperated in design of micro-hydropower plants and turbines. He carried out various short-term consultancy missic for development aid projects in Africa, the Caribbean and Columbia.

Die Deutsche Bibliothek – CIP-Einheitsaufnahme

Manual on induction motors used as generators : a publication of Deutsches Zentrum für Entwicklungstechnologien – GATE, a division of the Deutsche Gesellschaft für Technische Zusammenarbeit (GTZ) GmbH / Jean-Marc Chapallaz ... – Braunschweig : Vieweg 1992
(MHPG series harnessing water power on a small scale ; Vol. 10)

NE: Chapallaz, Jean-Marc; Deutsches Zentrum für Entwicklungstechnologien <Eschborn>; Mini Hydro Power Group: MHPG series harnessing ...

ISBN 978-3-528-02068-2 ISBN 978-3-663-14044-3 (eBook)
DOI 10.1007/978-3-663-14044-3

Originally published by Deutsche Gesellschaft für Technische Zusammenarbeit (GTZ) GmbH, Eschborn in 1992.

0. PREFACE

This handbook sets out to indicate how an induction motor could be used as generator, in the following abbreviated to IMAG. In addition to providing a practical selection method for IMAGs, the handbook covers all aspects related to the operation and control of the machine, safety and protection of the electrical installation and its economic advantage in comparison with synchronous generators
generators.

The contents of this handbook should be helpful to engineers and technicians engaged in micro-hydropower projects. It is not only intended for the hydropower engineer specialized in electrical engineering but rather for all those faced with the problem of electricity generation at a reasonable cost. The formal theory of electrical engineering used in the book has been simplified in order that the non-specialized mechanical, civil, rural or agricultural engineer should be able to follow all aspects covered by the book and to undertake the necessary computations without difficulty.
Moreover, Appendix A provides a short introduction into the basics of electrical engineering. It should be studied previously to the main text by those not being familiar with electricity generation and distribution.

S.I. units have been used throughout the book and standard symbols for physical properties employed.

The authors would like to express their special thanks to Prof. Simond of the Electrical Department at the Federal Institute of Technology in Lausanne (Switzerland) for his valuable support and assistance. Acknowledgments are also due to the following individuals and firms for their helpful criticism and supply of data:

- Messrs. Alex Arter and Jorge Senn, SKAT, Switzerland

- Mr. Berger, Geb. Meier AG, manufacturer of electrical machinery, Zurich

- Mr. H. Brüniger, Electrical Engineering, Chur, Switzerland

- Mr. B. Oettli, former engineer with BYS, Nepal

- Mr. Schoch, SRE electricity company, Clarens, Switzerland

- Mr. Schopfer, Federal Inspectorate of Electrical Power Supply and Distribution, Lausanne, Switzerland

- Mr. Nigel Smith, Trent Polytechnic; Nottingham, UK

HOW TO USE THE MANUAL

Chapter 2 introduces the basic principles of synchronous and induction generators (or IMAGs), their advantages and drawbacks. The handbook does not cover the generator as an isolated part of a micro-hydropower plant (MHP) but stresses the interdependence of the hydraulic, generating and consumer system. Various combinations in which an IMAG might be installed are examined: operation in parallel to a large grid, in parallel to other MHPs (minigrid) or as a stand-alone electricity generating unit.
To help the reader understand the working principle of an induction machine, a simplified theory including the differences between the (normal) motor operation and the generating mode is presented in Chapter 3.
Chapter 4 proposes a practical selection procedure with diagrams based on test results of standard induction motors.
Chapter 5 deals with the operation and control of an MHP equipped with an IMAG while Chapter 6 provides a general overview of safety and protection measures required in small-scale electricity generation and distribution schemes.
A simple method of dealing with economics in conjunction with energy production is proposed in Chapter 7. Chapter 8 includes a number of worked examples which show step by step the procedure of selecting an IMAG an comparing it with conventional generators.

The flowchart of Figure 0.1 below provides an overview of the contents of the manual; it should be seen as a guide to readers of different levels and interests on how to use the handbook in its most effective way.

FIGURE 0.1 :
Guide through the handbook

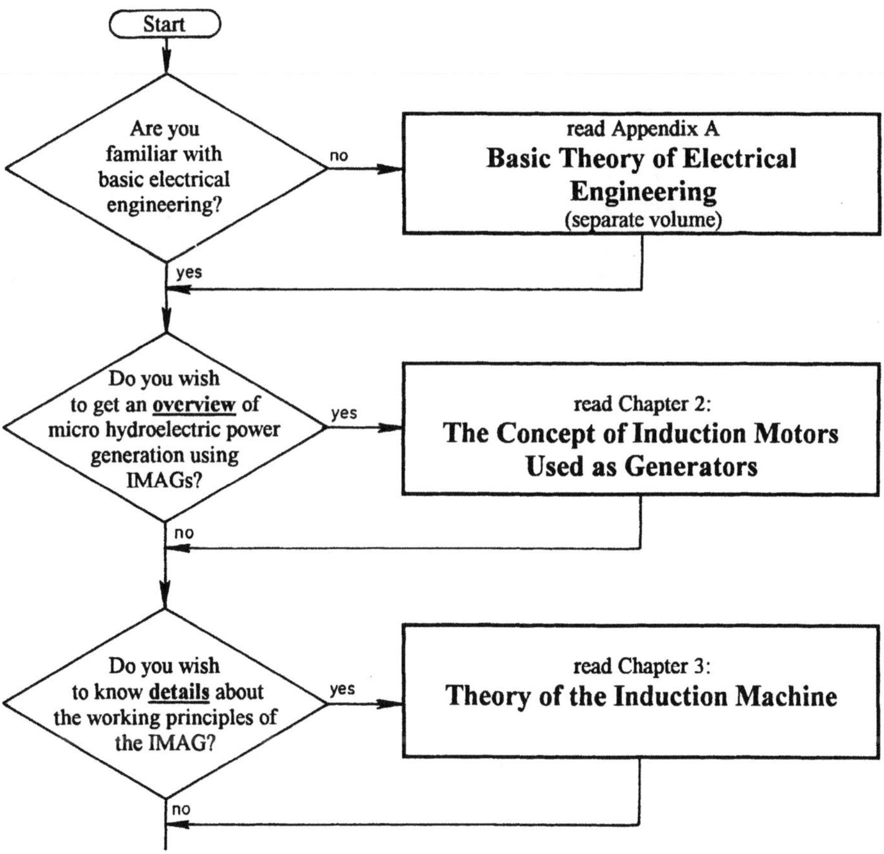

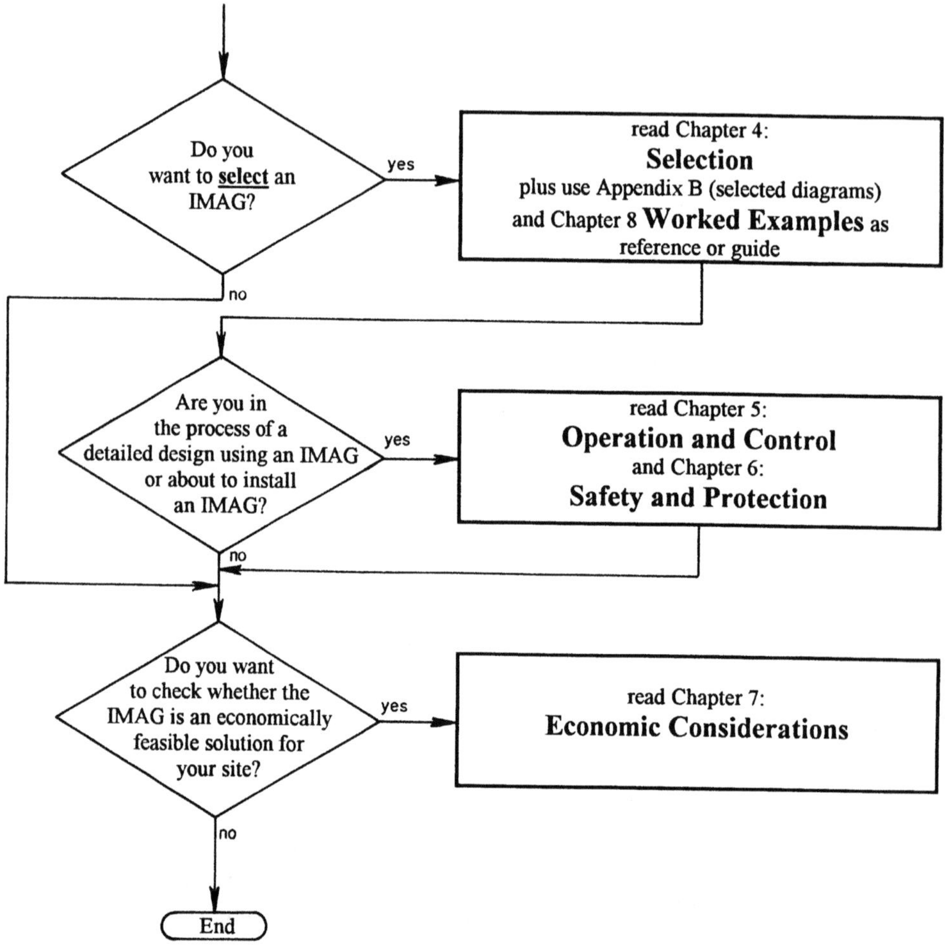
Do you
want to select an
IMAG?
yes
read Chapter 4:
Selection
plus use Appendix B (selected diagrams)
and Chapter 8 Worked Examples as
reference or guide
no
Are you in
the process of a
detailed design using an IMAG
or about to install
an IMAG?
yes
read Chapter 5:
Operation and Control
and Chapter 6:
Safety and Protection
no
Do you want
to check whether the
IMAG is an economically
feasible solution for
your site?
yes
read Chapter 7:
Economic Considerations
no
End

TABLE OF CONTENTS

Appendices:

Appendix A:
BASIC THEORY OF ELECTRICAL ENGINEERING

LIST OF SYMBOLS

Symbol	*S.I. Unit*	*Quantity*
A	m^2	Area
a	-	Inflation rate
B	$T = Wb/m^2$	Magnetic induction or magnetic flux density
C	F = A s / V	Capacitance
g	m/s^2	Gravitational acceleration
E	Ws (or kWh)	Electric energy
F	N	Force
f	Hz	Frequency
H	A/m	Magnetic field strength
H	m	Turbine pressure head (liquid column)
I	A	Current
I_μ	A	Magnetizing current
i	-	Interest rate
l	m	Length
L	H = Vs/A	Inductance
N	-	Number of turns of a coil / North pole / Neutral
n	1/min, rpm	Rotational speed
n_s	1/min, rpm	Synchronous speed
P	W	(Active) power
p	-	Number of pole pairs
R	Ω	Ohmic resistance
S	VA	Apparent power / South pole
s		Slip
t	s	Time
T	°C	Temperature
T	Nm	Torque
Q	Var	Reactive power
Q	m^3 / s	Flow, discharge
U	V	Voltage
v	m / s	Velocity
X	Ω	Reactance
X_m	Ω	Magnetizing reactance
X_σ	Ω	Leakage reactance
Z	Ω	Impedance

Symbol	*SI-Unit*	*Quantity*
α	°	Angle
Φ	Wb = V s	Magnetic flux
ϕ	°	Phase angle
η	-	Efficiency
μ	Vs/Am	Permeability of a magnetic circuit
ρ	$\Omega mm^2/m$	Resistivity of conductor material
ρ	kg/m^3	Density
Ω	rad/s	Angular speed
τ	s	Time constant
Δ	-	Delta connection
Y	-	Star (Wye) connection

Subscripts	*Meaning*
0	No load
1	Phase 1 / Primary winding (transformer)
2	Phase 2 / Secondary winding (transformer)
3	Phase 3
C	Capacitive
Cu	Copper
eff	Effective values
el	Electric
Fe	Iron
i	Instantaneous value
L	Inductive
n	Nominal value
g	Generator mode
m	Motor operation
m	Magnetizing
mec	Mechanical
r	Rotor
s	Stator
st	Starting

LIST OF ABBREVIATIONS

AC	Alternating current
DC	Direct current
AVR	Automatic voltage regulator (of synchronous generators)
IMAG	Induction motor as generator
IGC	Induction generator controller
ELC	Electronic load controller
IP	International protection
LV	Low voltage
MV	Medium voltage
HV	High voltage
MHP	Micro hydropower / micro-hydropower plant
O&M	Operation and maintenance
PAT	Pump as turbine
PE	Protective earthing
RMS	Root mean square values (effective values)
RF	Recovery factor (economics)
rpm	Revolutions per minute
C-2C	System of connecting capacitors on a three-phase IMAG serving a single phase load

1. INTRODUCTION AND OVERVIEW

Many small-scale hydroelectric power projects have failed already in the design stage because of insufficient means to purchase the electrical equipment (generator and controls). Unlike the mechanical (turbine) and the civil engineering components of an MHP, generators are still rarely manufactured in developing countries and importing such equipment is often beyond the capacity of a local project.

In MHPs of the lower output range, DC generators from cars and lorries have been used because of their ready availability. However, results have not been very promising due to the inconvenience of the low voltage generated (requiring special appliances), the limited life-span of the machines and difficulties with battery charging.

Another alternative to the conventional but often (too) expensive synchronous generator has only recently made its way into the field of micro-hydropower: the use of standard Induction Motors As Generators (**IMAG**). Astonishingly enough the possibility of using the induction motor as generator was first mentioned as early as 1893 but the standard non-modified motor has not found a widespread application as a generator in hydropower.

Interest in this solution has grown over the last decade in conjunction with the electrification of single estates and small villages in developing countries. The main advantage of the IMAG is its ready availability. Induction motors are widely used in industry and agriculture in practically any country and therefore tend to be less expensive than conventional generators, especially if they are locally manufactured or assembled. The advantage of cost and availability is most appreciable for machines between 0.5 and 50 kW while above 100 kW the conventional synchronous generator will usually be competitive. Furthermore, the simple and rugged construction of the induction machine ensures reliability and low maintenance costs.

The IMAG cannot generate magnetizing or reactive power by itself; to establish its magnetic field, the IMAG requires reactive power to be supplied to it either by synchronous generators working in parallel, or by capacitors in the case of stand-alone electricity generation. Capacitors add to the costs of the equipment and they may also fail.

Generally, the three-phase squirrel-cage induction motor is the preferred machine for generator operation. However, not all of the various designs and makes available on the market show good performance when generating although they may perform well as motors. The problem of how to distinguish the suitable from the unsuitable IMAGs cannot easily be solved. The engineer in the field may not have the possibility to carry out tests previous to installing the IMAG in a MHP, especially not with machines of higher ratings (> 3kW). Literature on the subject and published test results are few; manufacturers of motors see rarely an interest to measure and publish test results of their standard machines in generator mode.

Most diagrams and figures presented in this manual are based on tests on small induction motors (power output 0.4 - 7.5 kW) run at the laboratory of the Federal Institute of Technology in Lausanne (EPFL), Switzerland. Experience with larger IMAGs has been gathered from MHP projects implemented over the last few years in Switzerland.

However, collected data on the performance of induction machines in generator mode are not yet sufficient to exclude design errors or poor performance of selected IMAGs. The selection diagrams presented in this manual show a relatively large dispersion which can only be improved through further tests or experience from the field.

Aware of these difficulties, the authors dedicated a relatively large part of their research work to the question of how to improve the performance of an IMAG if the desired output is not obtained. The result of this work is presented in Chapter 5, Operation and Control, and includes measures to alter the connection of the windings (in parallel operation) or the voltage and frequency level in stand-alone electricity generation.

A number of additional problems concerning the application of IMAGs have been encountered in the course of the preparation of this manual but treating these in details would have been beyond the scope of this first edition. It is intended to present such problems in separate publications covering topics such as battery charging in conjunction with IMAGs, balancing a single-phase load on a three-phase IMAG or generator performance curves of selected motors available in developing countries.

Some of the disadvantages of the IMAG, such as the relative uncertainty of its performance in generator mode or the low power factor, can be improved by rewinding the stator of the machine. Specialized workshops in Europe engaged in hydroelectric power generation have been using this technique of producing purpose-wound induction generators from standard motors for many years and have acquired large experience. Electrical workshops in developing countries will certainly apply the same technique once a market for induction generators has evolved. The IMAG as a standard non-modified induction motor must therefore be seen as only a first step in the development of small-scale hydroelectric power in developing countries.

This handbook should serve as a sound foundation for the application of induction machines in hydropower schemes. With the combined experience of the readers of this manual, a more comprehensive reference could be established in future publications.
A feedback from the readers of this handbook will therefore be highly appreciated.

2. THE CONCEPT OF INDUCTION MOTORS USED AS GENERATORS

2.1 Introduction

The principal decision whether to use an induction or a synchronous generator for electricity generation depends on a large number of parameters. Not only the technical properties of the machines but also the systems into which the generators will supply energy have to be considered. Depending on whether the generating unit of a small hydropower plant covers the needs in electricity of a single estate or a small village, or whether it is linked to a larger national or regional grid, we may speak of **stand-alone and parallel operation** respectively. Obviously, the two modes of operation have different requirements on the generating unit (which generally comprises turbine, generator, governor and the electrical safety and control equipment).

For **parallel operation**, governing is usually simpler or not required at all since voltage and frequency are given by the grid. No investigations into the distribution and consumer system are required of the micro-hydropower designer in order to decide on the composition of the generating unit (choice of induction or synchronous generator). However, requirements by the national electricity utility on safety and control equipment might be fairly demanding.

For most of the micro-hydropower plants in developing countries, parallel operation will be excluded still for many years to come simply by the lack of integrated grid systems or of secured continuous supply of electricity by existing mains. In the scope of this manual we will therefore concentrate more on free-standing electricity generation. MHP schemes attempting to work in parallel with a large grid will have to comply with the rules and regulations set out by the electricity utility concerned. This manual can therefore only provide some general information on parallel operation.

Micro-hydropower plants working as **stand-alone electricity generating units** must maintain voltage and frequency of the generated electricity constant , independent of varying electrical loads or varying flow of water to the turbine. Until recently, synchronous generators were therefore regarded as the only option for stand-alone electricity generation of a certain size. The development and mass production of solid-state controls now provides reliable and economically feasible solutions for controlling the stand-alone induction generator. Nevertheless, designing an electricity generating plant still requires investigations into all the system components of electricity generation and consumption and their interdependence, despite the promising new technologies. In the following sections, we will give an overview of the system components and the parameters which have to be studied in order to decide whether an IMAG may be a convenient solution for a certain MHP.

2.2 Induction versus Synchronous Generator

2.2.1 Working Principle of a Synchronous Generator

Electricity generators convert mechanical energy into electricity. The synchronous generator is a rotating machine generating single or 3-phase alternating current with a frequency proportional to its rotational speed. DC generators which provide a means to store energy in batteries are not considered in this manual.

The synchronous generator consists of a stationary member, the stator, comprising windings, and of a rotating magnetic wheel, called rotor (see Figure 2.1 below). The rotor may be a permanent magnet (for very small output) or more conveniently an electromagnet whose coils or windings are fed by an (external) DC source called excitation. The rotating magnetic field (created by the rotor) induces voltage and current in the stator windings. To generate voltage at constant frequency, the speed of the rotor must be kept at synchronous speed (e.g. 3000 rpm for a 2-pole generator for 50 Hz). The voltage of the generated electricity depends on the excitation current. To increase the output voltage of the generator, the magnetic field of the rotor must be stronger, which can be achieved by an increase of the excitation current.

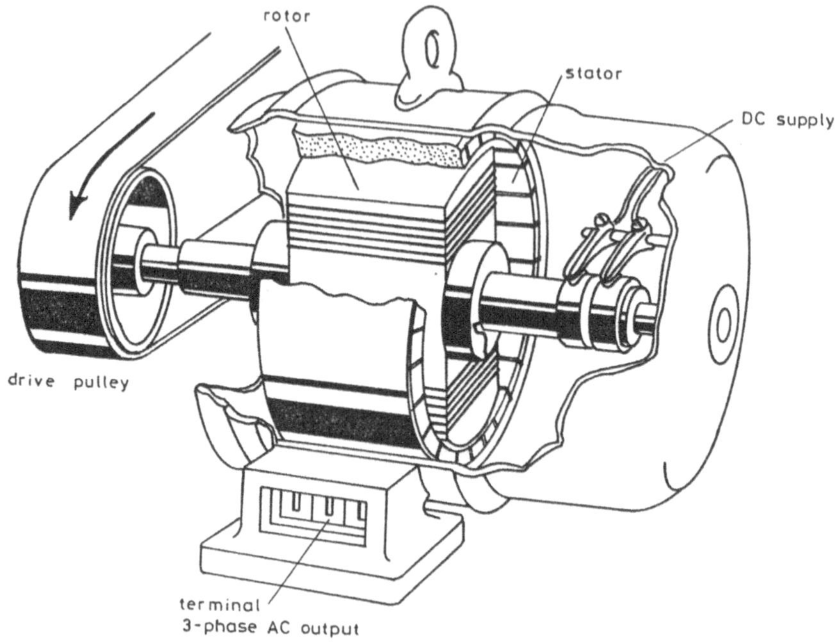

FIGURE 2.1 :
Schematic representation of a synchronous generator

There are two designs of synchronous generators available on the market: the brushed and the brushless generator. They differ from each other mainly in the way they generate and supply the excitation current.
The brushed generator employs a small DC generator, called exciter, mounted on the main shaft (or an external rectifier) to generate the excitation current. This current is supplied to the rotor of the synchronous machine by slip rings and carbon brushes. These need to be replaced periodically and the brushed generator therefore requires a continuous maintenance effort.
Today, the brushless type is the most commonly used synchronous machine in the micro-hydropower range. It dispenses with slip rings and brushes and uses a small AC generator mounted on the rotor shaft to generate the excitation current. This AC current is rectified on the rotor itself by a solid-state controller (rotating rectifier) so as to supply the rotor windings with the required DC current. The absence of brushes and slip rings makes the unit more reliable and virtually maintenance-free.

When an inductive load is put on the synchronous generator, the terminal voltage will drop; for a capacitive load, it will increase. In order to maintain constant voltage at the machine terminals, a so-called automatic voltage regulator (AVR) controls the excitation current, i.e. the rotor field flux is increased (by increasing the field excitation current) when voltage drops and vice versa. Synchronous generators may be operated in the over excited range to provide reactive power for induction motors and other inductive loads.
Figure 2.2 shows a brushless synchronous generator with an automatic voltage regulator.

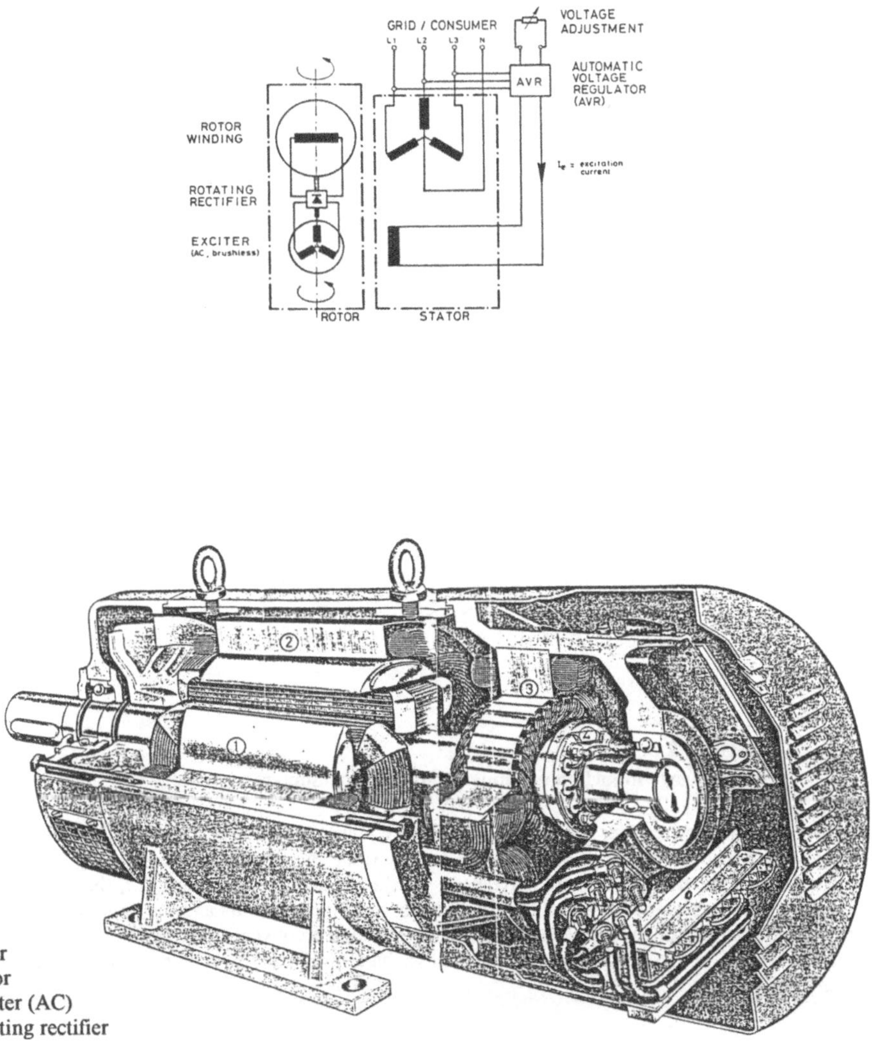

1 Rotor
2 Stator
3 Exciter (AC)
4 Rotating rectifier

FIGURE 2.2 :
Brushless synchronous generator with automatic voltage regulator (AVR)
(Source: Geb. Meier, Zurich / AvK, Germany)

2.2.2 Working Principle of an Induction Machine

A distinctive property of the induction machine is its rugged and simple construction. It consists basically of a stationary winding, called stator, enclosed by the machine frame and a rotor with a short-circuited winding.
Placing a three-phase AC current on the terminals of the three-phase stator winding, creates a rotating magnetic field in the machine which rotates at a speed, called synchronous speed n_s, depending on the supply frequency and the number of pole pairs of the induction machine ($f = p\, n_s$). The rotating field flux cuts the short-circuited rotor winding (or conductor bars in the case of the squirrel-cage rotor) where it induces voltage and current which in turn produce torque on the rotor. The rotor must always rotate below or above synchronous speed, i.e at a slip, otherwise there is no cutting of flux by the rotor conductors and hence, no torque developed. The induction machine operates as a motor when running below synchronous speed and as a generator when rotor speed is above synchronous speed. Therefore, any induction motor may also be used as a generator simply by driving it above synchronous speed.

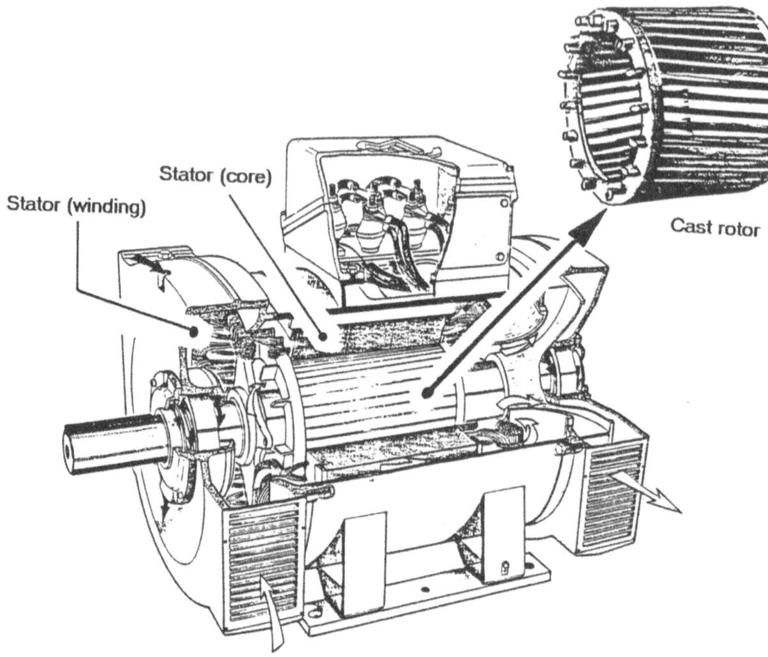

FIGURE 2.3 :
Squirrel-cage induction machine (source: Leroy Somer)

The induction generator requires magnetizing or reactive power to establish its magnetic field. In parallel operation, reactive power can be drawn from the grid to which the IMAG supplies active power; in stand-alone operation, a battery of capacitors is required. Unlike the synchronous generator, the power factor of the induction machine cannot be adjusted; it is fixed by machine constants and the operating speed (slip). More details on the working principle of the induction machine are given in Chapter 3. Figure 2.4 shows a schematic diagram of the induction machine with a battery of capacitors connected for free-standing electricity generation.

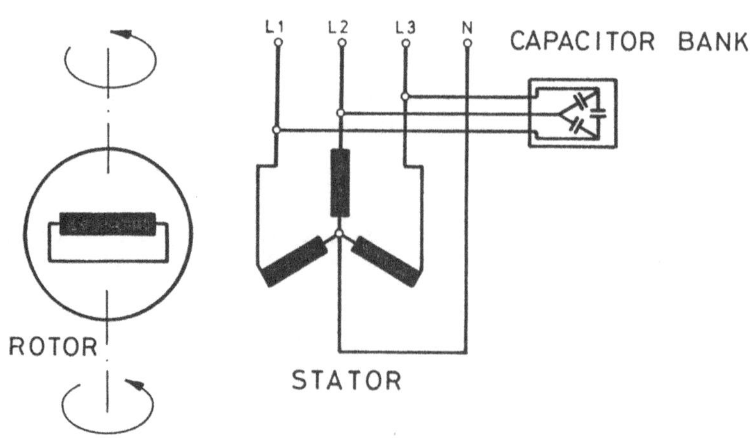

FIGURE 2.4 :
Schematic diagram of an induction machine with capacitor bank

2.3 System Components of an MHP with Stand-alone Electricity Generation

Basically, we can distinguish four main components of an MHP generating electricity which may all have in one form or other an effect on the selection of the generator (see Figure 2.5):

- The ***hydraulic system*** comprising the intake, sand trap, head race channel or conduit, forebay or surge tank, penstock , turbine and tail race

- The ***electricity generating system*** (in short: electrical equipment) covering the generator, the monitoring and protection equipment and the switchgear (+ possibly a transformer)

- The ***consumer system*** including distribution lines and the appliances or electrical loads (lighting, radios, TVs, stoves or cookers, heaters, motors, etc.)

- These three systems are interrelated by the ***governing system*** which matches the power demand with the supply of the generator. The governing system comprises two functions:

  - control of the generated frequency by a turbine governor (or a load controller)

  - control of the voltage of the system by an AVR (automatic voltage regulator) or a load controller

  These two factors properly controlled within the permissible range of variation determine the quality of the generated electricity.

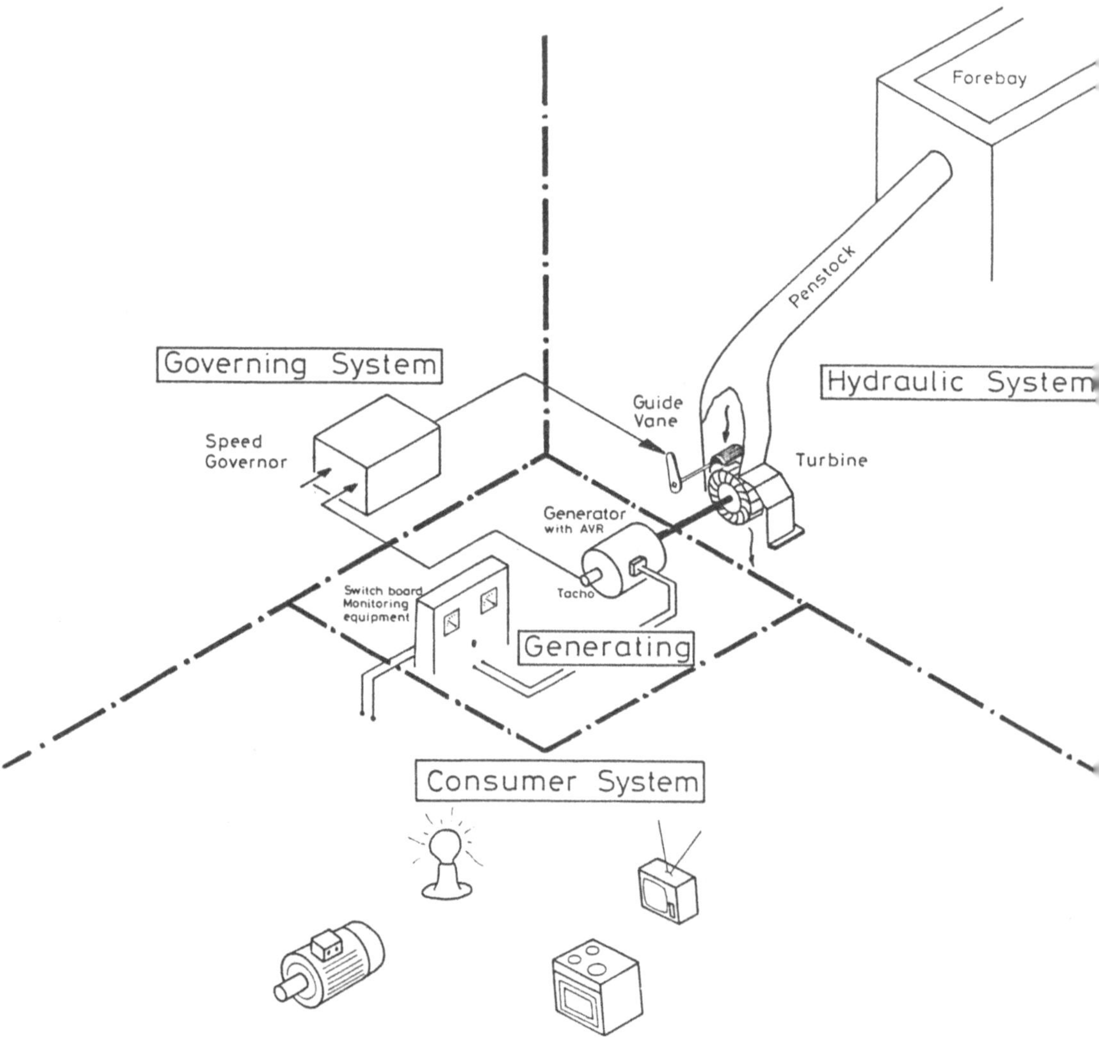

FIGURE 2.5 :
System components of an MHP for stand-alone electricity generation

2.3.1 Possible Applications of Induction Motors as Generators in MHPs

Figure 2.6 shows various fields of applications and combinations of induction machines used as generators in micro-hydropower plants. Obviously, an induction machine may be used in virtually all possible applications of the micro-hydropower field. However, some applications or combinations are more favourable than others, either because of technical or economic constraints. In the following sections, we will give an overview of the advantages and disadvantages of the induction motor as generator in various applications mainly in comparison with the synchronous generator.

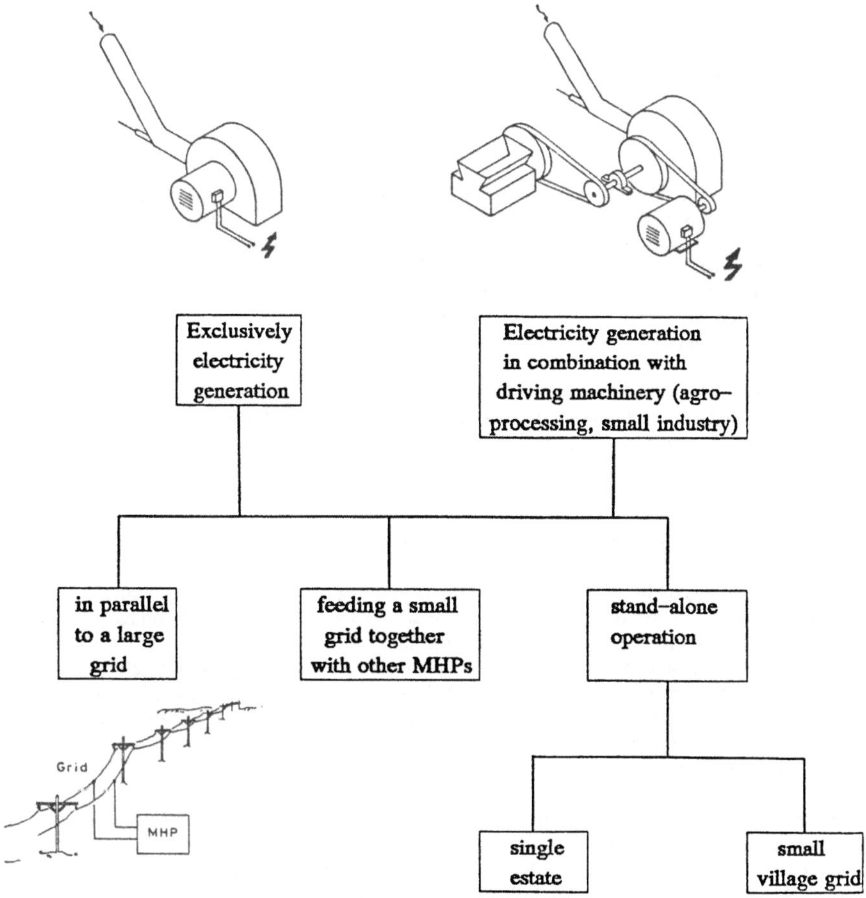

FIGURE 2.6 :
Possible applications of induction motors as generators in micro-hydropower plants

2.3.2 The Hydraulic System and its Effect on the Choice of the Generator

- Part-load efficiency

  Micro-hydropower schemes can seldom operate at nominal output all year round because of insufficient flow during the dry season. Turbines and generators will therefore operate at part load for long periods.

  As a general rule, generators are usually slightly overrated (i.e. a machine with a rating above the estimated maximum turbine output is selected) in order to avoid overloading during operation and also because the exact rating required may not be available as a standard-size machine. The safety margin adopted for IMAGs is usually somewhat higher than for synchronous machines (see Chapter 4: Selection of IMAGs). Since efficiency of the generator drops for any output below the nominal value, the overrated IMAG may have lower efficiency at part load than a comparable synchronous generator.

- Overspeed

The overspeed capability of a generator is an important criterion for hydropower plants since the generating unit will attain runaway speed (up to twice nominal speed and more) when load is rejected. For induction machines of 4 poles and more, manufacturers can usually guarantee a maximum speed of twice the nominal speed. Two-pole motors, however, have more limited overspeed capabilities (1.3 to 1.8 or occasionally 2.0 times synchronous speed). Additionally, these machines may also have slightly shorter life-span (higher wear due to high nominal speed). The use of two-pole motors as generators is therefore not recommended.

Synchronous generators are usually designed to stand runaway speeds of diesel engines, i.e. 1.2 to 1.3 times nominal speed. Higher overspeed capabilities required for turbine prime movers may only be available upon request (incurring an extra charge).

2.3.3 The Consumer System and its Effect on the Choice of the Generator

The quality of the generated electricity depends on the requirements of the end-use appliances. The different electrical loads of a scheme are designed to operate under constant frequency and voltage. Deviating from these nominal values might affect the lifetime of some applications while others might not work properly. Many modern household appliances (lamps, TVs, computers, absorber-type refrigerators, etc) can now operate at voltage variations of 230 V -15% and +10% and at both 50 and 60 Hz without any damaging effect. Generally, motors and transformers, i.e. inductive loads, are the most sensitive loads. A drop in frequency diminishes the reactance of the transformer or motor coils allowing more current to flow which may finally cause windings to burn.

Generating units using synchronous generators (with AVR and speed governor or load controller) usually provide a high quality of electricity with only small voltage and frequency variations. Induction motors used as generators are, in todays most common applications, not able to maintain frequency and voltage variations within the same range. In the following section we will introduce some of the frequently used governing and control systems and their corresponding voltage and frequency variations.

2.3.4 Governors and Controllers Required by Synchronous or Induction Generators

Stand-alone Electricity Generation

Electricity generation in isolated plants requires constant speed of the turbine/generator unit in order to maintain constant frequency and voltage. When switching on electrical loads, the speed of the generating unit will drop until a new equilibrium between generated and consumed power is reached. Such changes of speed and the subsequent voltage and frequency variations of the generated electricity are undesired from the point of view of the end-use appliances (see above).

- **Flow-Control**

The speed of the generating unit can be kept constant if, upon an increase of the electrical load, a corresponding increase of turbine output, i.e water flow through the turbine, is brought about. Or, in other words: generated power matches at any instant the consumed power due to progressive flow control. Note that in isolated plants, the available output of the generating unit must at any instant be higher than the total power demand of simultaneously connected consumers or the system will break down.
Figure 2.5 above shows a flow-control governor acting on the turbine guide vane. Flow-control governors exist in various designs: e.g. mechanical (oil-hydraulic) or electro-mechanical (using linear actuators). However, flow-control governors are usually of a fairly complex design (precision engineering) and require good maintenance; additionally, they are seldom manufactured locally and therefore tend to be very costly for small hydropower projects in developing countries. Furthermore, flow-control governors are more accurate in their performance than is usually required in rural areas. For these reasons, speed governing by means of flow control is seldom used in conjunction with IMAGs since the cost advantage of the generator would by far be ruled out by the high costs of a mechanical speed governor.

- **Load control**

An alternative to speed governing on the hydraulic system is the control on the generator output, called load control. In its simplest form it incorporates a consumer system of constant loads requiring no governor at all: the generating unit continuously operates at rated power which is at any instant absorbed by the consumers. When disconnecting an electrical load, another one of similar magnitude must be switched on immediately in order to keep generated and consumed power in equilibrium.
These adjustments of loads could also be done automatically by an electronic device called electronic load controller (ELC, for synchronous generators) or induction generator controller (IGC). These devices switch any power not consumed by the regular circuit into a ballast load (resistors, heating elements); hence, generator output power and consumed power always match.

Table 2.7 below gives an overview of possible speed governing options for both IMAGs and synchronous generators. Approximate frequency and voltage variations of the different systems are also indicated; these may be used as a general selection criterion; i.e. a certain option, e.g. an unregulated IMAG, may only be applicable if the planned end-use appliances of the isolated grid will not require a better quality of electricity.

TABLE 2.7 :
Overview of speed governing options for IMAGs and synchronous generators and their approximate range of voltage and frequency variations

Speed Governing by means of:	**IMAG** Governor/Controller	Voltage and Frequency Variations	**Synchronous generator with AVR incorporated** Governor/Controller	Voltage and Frequency Variations
Un-regulated: a) constant load	Only possible for constant loads (no motors etc.)	$\Delta U = \pm 15\%$ $\Delta f = \pm 10\%$	Most of todays synchronous generators are standard-equiped with AVR, hence the completely unregulated synchronous generator does practically not exist	
b) variable load	unregulated IMAGs will speed up when loads are partly disconnected and high overvoltage will occur: **dangerous,not applicable**	$\Delta U = 80\%$ - 160% $\Delta f = 80\%$ - 160%		
Hand-regulated (flow control)	speed control acting on turbine opening (flow) is not sufficient for an induction generator, additional voltage control is necessary **voltage** may be controlled by using a battery of **stepped capacitors** which are switched on and off by over/undervoltage relays according to the power requirements of the load	$\Delta U = \pm 15\%$ $\Delta f = >10\%$ (depending on operator)	operator increases flow through turbine when load increases; **AVR** regulates voltage	$\Delta U = \pm 2\%$ $\Delta f = >10\%$ (depending on operator)
Flow control	speed control acting on turbine opening (flow) is not sufficient; additional voltage control is necessary **Two governors/controllers are (too) expensive** (electro-mechanical speed governor sensing voltage and acting on the turbine gate opening (using a linear actuator) could be an interesting option: has been applied on diesel gen-sets but not on turbines so far)	$\Delta U = \pm 2\%$ $\Delta f = 5\%$ - 10%	**mechanical speed governor** senses frequency and acts on the turbine gate opening; AVR regulates voltage	$\Delta U = \pm 2\%$ $\Delta f = \pm 0\%$ - 6%
Load control	surplus power is diverted to a ballast load so that turbine power and speed are maintained close to their rated values: the **induction generator controller IGC** senses **voltage variations** (through solid state controllers) to activate the ballast loads	$\Delta U = \pm 2\%$ $\Delta f = 5\%$ - 10% (depending on load power factor)	surplus power is diverted to a ballast load so that turbine power and speed are maintained close to their rated values: the **electronic load controller ELC** senses **frequency variations** (through solid state controllers) to activate the ballast loads; the AVR regulates voltage	$\Delta U = \pm 2\%$ $\Delta f = \pm 0\%$ - 6%

Note that the indicated voltage and frequency variations do not apply to sudden changes of load. Transient voltage or frequency deviations are usually higher (e.g. when large motors are started). Protection of the other consumers during transient periods might have to be considered.

Parallel Operation
Electricity generation in parallel to a large grid does not need a speed governor or a voltage controller since the grid dictates voltage and frequency of the generating unit. An induction motor can make a very suitable generator for parallel operation as it is not only less expensive than a synchronous generator, it neither requires synchronizing devices.
The reactive power demand of the induction generator from the grid may be reduced or compensated by connecting a battery of capacitors according to the power factor requirements of the electricity utility. Compensation of reactive power demand will reduce line current, hence transmission losses.

To make maximum use of the available water, **manual or automatic water level control** is advisable. At large stream flows, the turbine gate is opened to its maximum while during the dry season, the guide vanes are adjusted in such a way that the water level in the forebay,i.e. turbine head, is maintained. Here, IMAGs may have a certain disadvantage in comparison with synchronous generators: at part load, efficiencies of IMAGs may be lower than those of synchronous generators.

The problem of governing especially for MHP applications using an IMAG will be treated in more details in Chapter 5, Operation and Control.

2.4 Pros and Cons of Induction Motors Used as Generators

The main advantage of the induction motor as generator does not arise from technical considerations as discussed so far. Interest in IMAGs for their application in small hydropower projects has grown because of the lower capital costs of induction motors as compared to any other generator and because of their widespread availability. The following figure compares the purchase price of an IMAG with that of a synchronous generator (brushless with automatic voltage regulator (AVR) in function of power output). The prices are valid for three-phase motors and generators available on the European market (prices ex. works 1990). The induction motor requires a battery of capacitors when operating as a stand-alone electricity generating unit. The figure reveals that:

- up to 10 kVA output, the capital costs for induction motor and capacitor bank is half the price of a corresponding synchronous generator

- up to 25 kVA, the economic advantage of the induction machine is still about a third of the total investment costs.

In developing countries, synchronous generators are not easily available and might have to be imported by the project; the economic advantage of the induction motor (locally manufactured or assembled) may be even greater.

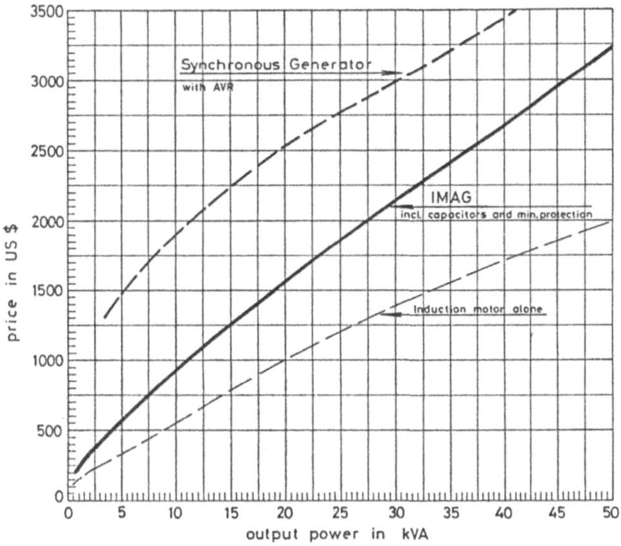

FIGURE 2.8 :
Cost comparison between induction motors used as generators and synchronous generators (4 pole machines, n_s = 1500 rpm; prices of the European market, ex works 1990)

The following table gives an overview of the advantages and disadvantages of the induction motor used as generator in comparison with a synchronous generator.

Advantages of the Induction Motor as Generator (IMAG)

Criteria	IMAG	Synchronous Generator
Investment costs	for lower power range (**up to 30 kVA**) considerable advantage	for medium power range (over 100 kVA) competitive because induction motors of this range no longer widely available
Availability of machines and spare parts	**readily available** due to widespread use of induction motors in agriculture and industry	not readily available in developing countries due to small market (=> may have to be imported)
Construction	rugged and simple, ensuring **reliability and low maintenance costs**	DC exciter, brushes and slip rings, insulated rotor windings plus AVR make the synchronous generator more liable to breakdowns and higher costs of maintenance and repair
Synchronizing for parallel operation	**no synchronization** required - main circuit breaker is simply closed near synchronous speed	requires synchronizing equipment (synchronoscope or lamps)
speed control governor for isolated operation	relatively **simple equipment** is sufficient in most cases; e.g. IGC which controls voltage since frequency will not vary excessively due to saturation of induction motors	synchronous generators require usually both voltage (AVR) and frequency control (e.g. ELC) in unattended stand-alone electricity generation
Overspeed capability	Motors of 4 and more poles usually stand twice nominal speed; 2-pole machines may have more limited overspeed capabilities	Overspeed capabilities required in conjunction with turbines only upon request

Disadvantages of the Induction Motor Used as Generator (IMAG)

Criteria	IMAG	Synchronous Generator
Magnetizing power	requires magnetizing power to be supplied to it either from grid or from capacitors (which are expensive and may fail)	can supply reactive power if required through adjustment of excitation current
Efficiency	generally lower than for synchronous generators; decrease for small machines; relatively low efficiency at part load	maintains high efficiency at part load
Power factor	progressively declining the lower the rated speed and the power rating of the machine; generally lower power factor than for purpose-wound induction generators	can be adjusted according to needs with excitation current
Inertia of rotating masses on load rejection	rotor of induction machines have low inertia and IMAGs sometimes require a flywheel where a synchronous machine could still do without (to limit acceleration of generating set and consequently to avoid water hammer problems in longer penstocks)	rotor inertias higher than for IMAGs
Quality of generated electricity in isolated operation	voltage and frequency variations higher (even with IGC) than for standard synchronous generator sets with comparable control equipment	AVR in conjunction with ELC or governor maintain voltage and frequency at a fairly high level
Selection of the machine	few support from manufacturers of induction motors; risk of design errors by MHP engineer (some fairly good motors may have very poor performance as generators and may cause problems, overheating)	less demanding selection procedure but compatibility of AVR with the selected governor must be verified (which is not a simple problem either)
Large motor starts	IMAG may de-excite (system breakdown)	Short-term overload capacity allows direct-on-line starts of large motors

2.5 Suitable IMAGs

The preferred machine to be used as a generator is the squirrel-cage motor with a simple rotor bar design (no double-cage rotors). The wound-rotor type may be used but is slightly less sturdy than the squirrel-cage design, is more expensive and not always readily available.
The machine should have high efficiency and a low full-load slip and low starting but high pull-out torque. Modifications which may improve performance in generating mode are frequently executed by specialized workshops and concern mainly the stator winding. The cost advantage of modified induction machines as compared with synchronous generators is, however, reduced. Moreover, a vast experience with electrical machines and suitable equipment is required when such modifications are attempted. In the course of this manual we will not treat the procedure of rewinding induction machines.

2.6 Conclusions

The standard squirrel-cage induction motor can be used very efficiently and without modifications as an induction generator in small hydropower schemes and power recovery systems up to about 50 kW in isolated systems and up to about 100 kW for parallel operation. Apart from its cost advantage in comparison with the synchronous generator it has a number of additional advantages such as simplified controls and less maintenance requirements. However, other aspects such as its relatively low part-load efficiency in generating mode and its fairly demanding design procedure must be considered.

3. THEORY OF THE THREE-PHASE INDUCTION MACHINE

3.1 General Design Features of the Induction Machine

The induction or asynchronous machine is the most commonly used AC machine. Like any electrical machine, it can be used both as a motor or a generator. The name induction machine relates to the fact that the stator winding creates a rotating magnetic field which induces alternating current into the short-circuited rotor winding.

The induction machine generally comprises two parts: the stationary part called stator and the rotating cylindrical part, referred to as rotor. Both stator and rotor iron are composed of laminations of a few tenth of mm in thickness cut from iron sheets. This is to minimize the iron losses (hysteresis and eddy currents, see Appendix A) produced by the alternating magnetic fields. The laminations of the stator are slotted at the inner surface to accommodate a three-phase winding. This winding comprises coils of several turns distributed over the stator periphery. The coils of each phase are connected in series. The desired number of poles and the related speed/frequency is attained by a corresponding spacing of the coils (pole pitch: 2-pole motor: 180° , 4-pole machine: 90° as in Figure 3.1). Figure 3.1 shows the stator of an induction machine with four poles (synchronous speed = 1500 rpm / 50 Hz, see also 3.2.1). The ends of each coil are led to the terminal box and may be connected either in star (Y) or in delta (Δ) to the three conductors of the network.

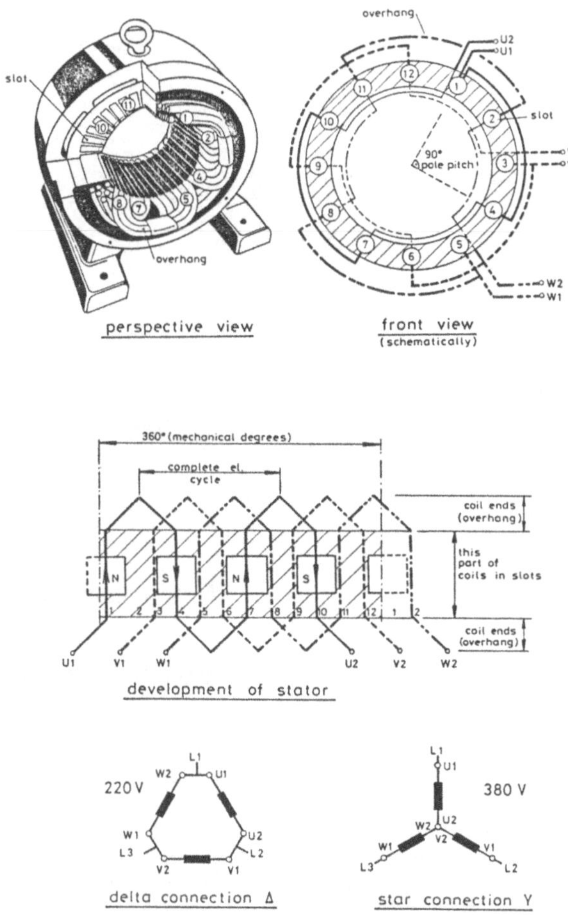

FIGURE 3.1 :
Stator of an induction machine with a four-pole winding

The rotor winding design varies depending on the need for torque or speed control. Two general categories can be distinguished: the squirrel-cage rotor and the wound-rotor type. However, the preferred induction machine used as generator is the squirrel-cage type. It consists of copper or aluminium bars embedded in the rotor slots. All the bars are short-circuited by two end rings as shown schematically in Figure 3.2. In small machines the complete rotor winding, including bars, end rings and fan are cast in place (directly into the slots of the rotor iron without insulation). The air gap between the stator and the rotor iron is relatively small (about 1 mm). There are no slip rings or carbon brushes required because the rotor winding is short-circuited internally on the rotor. The squirrel-cage machine is therefore virtually maintenance-free.

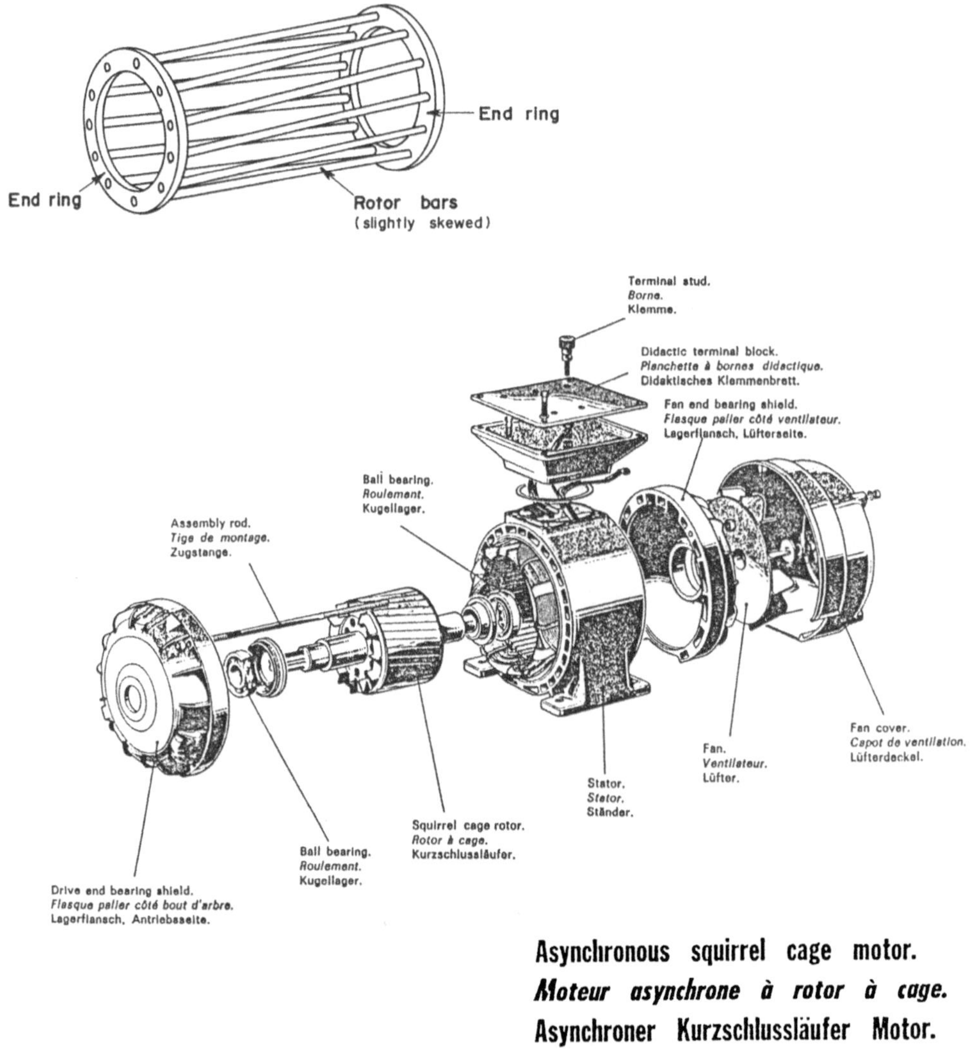

FIGURE 3.2 :
Asynchronous squirrel-cage machine (Source: Leroy Somer)

The second type of rotor construction is the wound rotor. The rotor has a three-phase winding similar to the stator and wound for the same number of poles as the stator winding. The rotor windings terminate in slip rings mounted on the rotor shaft. Brushes ride on the slip rings and, during the starting period (for motor operation only) are connected externally to a resistor bank which can be shorted out as the motor comes up to speed. Thus, this rotor construction has the advantage of varying resistance which may be required to reduce starting current, to improve starting torque or to alter the operational speed. Since these features are not relevant for generator operation, the wound-rotor motor is seldom used as generator mainly due to its higher price. It will therefore not be further discussed in this manual although, from a technical point of view, the wound-rotor motor may perform perfectly well in generator mode

3.2 Working Principle of an Induction Machine

3.2.1 The Rotating Magnetic Field

In motor operation, the stator winding is supplied with a three-phase voltage. Currents will flow in each phase of the winding. These currents will be displaced from each other by 120°. Consider a simplified two-pole three-phase stator winding as shown in Figure 3.3 . The current flowing through each phase will create a magnetic field with a polarity in the direction of its magnetic axis (right-hand-rule, see Appendix A) which is displaced from the coil by 90°. Because of the alternating waveform of the applied current, the magnetomotive force of each phase is also oscillating between a positive and a negative maximum. Although each of the three vectors of the magnetomotive force is stationary, the resultant magnetic field (= the sum of three components) Φ_s seems to rotate. This is illustrated in Figure 3.3. For the four instants of time shown, the resultant magnetomotive force is constant in magnitude and revolves the distance covered by the coil spacing (180° in a 2-pole machine) in one-half cycle.

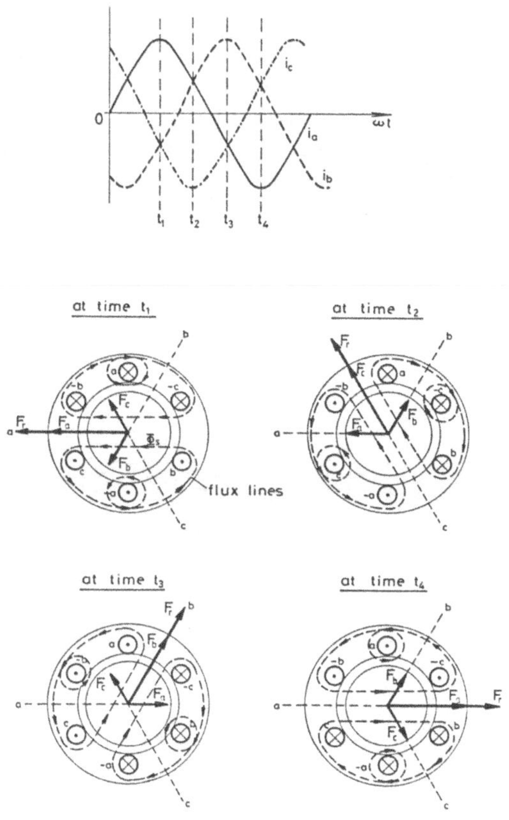

FIGURE 3.3 :
Illustration of the rotating magnetic field for four time instants of a two-pole three-phase stator winding

Figure 3.3 uses the following convention: (+) indicates a stator current directed into the page while (°) refers to a current flowing out of the page. At time $t = t_1$, the current i_a in phase a - a' is positive maximum, and i_b and i_c are negative and equal to one half of their respective maxima. Therefore, the magnetomotive forces or fields due to these currents will have corresponding magnitudes. Their directions are determined by the right-hand rule. Combining these three components graphically yields a resultant magnetomotive force which is 1.5 times the maximum of one phase.

If a magnetic needle is mounted onto the shaft of the machine, it rotates at the same speed as the magnetic field: in the case of the two-pole motor connected to a 50 Hz grid the speed of the rotating field is 3000 rpm. If the number of poles are increased, say 4 poles, the needle rotates only half the distance as before during one cycle of the supply frequency. This implies that the speed of the rotating field is inversely proportional to the number of pole pairs and proportional to the frequency of the supply voltage:

$$\boxed{n_S = \frac{f}{p}} \tag{3.1}$$

where n_S = synchronous speed of the rotating magnetic field in s^{-1}
f = frequency of the stator current in Hz (s^{-1})
p = number of stator pole pairs

Note that by reversing the phase sequence of the supply, by interchanging any of two currents, the field will rotate in the opposite direction. In practice, this is the procedure followed to reverse the direction of rotation of an induction motor (by interchanging any two of the motor terminals).

3.2.2 Slip and Rotor Speed

In the induction machine, the rotating magnetic field Φ_s created by the stator, cuts the rotor bars (rotor at standstill) and induces voltage into the rotor bars. When the rotor is at standstill, the frequency of this induced voltage is the same as the rotating magnetic field frequency. Due to the short-circuited rotor bars (by means of the end rings), a current, the induced current, will flow through the rotor. This current produces a rotating flux wave Φ_r similar to the one of the stator. Figure 3.4 below presents a simple induction machine with only one phase shown. The stator and the rotor fields interfere with each other and the resulting field creates a force couple or torque on the rotor bars. The torque on the rotor is directed in such a way that the rotor will turn in the direction of the rotating stator field. Figuratively, we can say that the rotor follows the main field Φ_s in order to catch up to it.

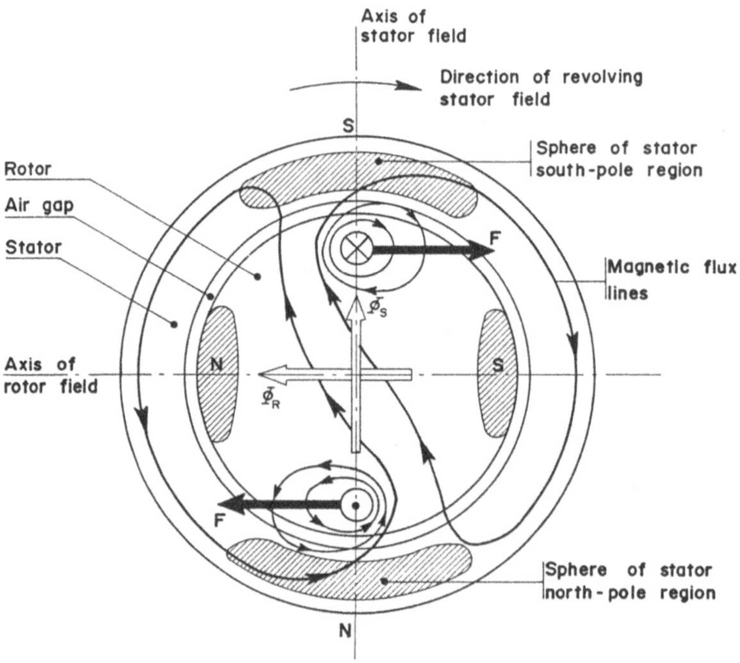

FIGURE 3.4 :
Development of torque in an induction motor (schematically for a single rotor coil)

However, there must always be a relative motion between the rotating stator field Φ_s and the rotor bars; if the rotor bars rotate at the same speed as the rotating field, i.e at synchronous speed n_s (see formula (3.1), the rotor conductors would no longer be cut by the flux lines. As a consequence, there would be no induced current and no corresponding rotor field Φ_r and hence no torque. That is why the induction machine is also called asynchronous machine as the rotor must rotate below (or above) synchronous speed to produce electro-magnetic torque. Even if the motor is run with the load uncoupled, a small torque is required to overcome friction and other losses of the machine and the rotor will rotate slightly below synchronous speed.
The difference between the synchronous speed n_s and the rotor speed n_r is called slip speed n_g and represents the speed of the rotating field viewed from the rotor. Relating n_g to the synchronous speed, we obtain the so-called slip s:

$$n_g = n_s - n_r \tag{3.2}$$

$$\boxed{s = \frac{n_s - n_r}{n_s}} \tag{3.3}$$

Slip s may be negative, i.e the rotor speed n_r is above synchronous speed; the machine operates as a generator driven by a turbine or a diesel engine and generates electric current in the stator windings.

Combining formulae 3.1 and 3.3 leads to:

$$\boxed{f = p \frac{n_r}{1 - s}} \tag{3.4}$$

3.2.3 Induced Voltage and Frequency

The frequency of the rotor voltage and current depends on the slip speed n_g . It equals the stator or grid frequency at standstill, because the rotational speed of the magnetic field is the same for both stator and rotor windings. With the rotor set in motion, the rotor frequency varies proportionally to n_g:

$$f_r = f \frac{n_g}{n_s} = s f \tag{3.5}$$

where f_r = rotor frequency in Hz (note this is the electrical frequency of the rotor current and voltage and not the rotor rotational speed)
f = grid frequency in Hz
s = slip

A similar formula applies for the rotor-induced voltage: At standstill when the rotor is at rest, the rotating magnetic field sweeps the rotor bars at its maximum rate and the induced voltage in the rotor will be maximum. As the slip decreases, the induced voltage decreases proportionally and reaches zero when the rotor rotates at synchronous speed. Hence, the induced voltage of the rotor writes:

$$U_r = s\, U_{r\,SC} \tag{3.6}$$

where $U_{r\,SC}$ is the rotor-induced voltage at standstill (blocked rotor).

3.2.4 The Equivalent Circuit

An induction machine resembles in many ways a transformer. As in a transformer, two (or more) electrical circuits, i.e. the stator windings and the rotor winding or bars, are coupled by means of a magnetic field so that electrical energy may be transmitted from one circuit to the other. The induction machine goes even further than the transformer since the rotor circuit is free to move continuously with respect to the stationary stator circuit and energy is transformed from electrical to mechanical (motor) or vice versa (generator operation).

Let us look at the rotor circuit first (on a per phase basis). The induced voltage U_r is only limited by the rotor impedance Z_r. This impedance is made up of two components: the rotor resistance R_r and the leakage reactance $X_{\sigma r}$. The leakage reactance represents that part of the magnetic field (created by the rotor current) which does not link with the stator winding. Since reactance is a function of frequency (Appendix A, 7.7.2: $X = 2 \pi f L$) the leakage reactance is proportional to the slip:

$X_{\sigma r} = s\, X_{\sigma r\, SC}$ where $X_{\sigma r\, SC}$ is the reactance at blocked rotor (SC stands for short circuit)

If we assume the leakage reactance and the resistance being connected in series, the impedance of the rotor becomes $Z_r = \sqrt{R_r^2 + s\, X_{\sigma r\, SC}^2}$ and the rotor current $I_r = U_r / Z_r$ or

$$\boxed{I_r = \frac{s\, U_{rSC}}{\sqrt{R_r^2 + s\, X_{\sigma r\, SC}^2}}} \qquad (3.7)$$

Figure 3.5 a) shows the rotor diagram comprising the resistance R_r and the reactance $X_{\sigma r}$.

If we divide both numerator and denominator of equation (6) by the slip we obtain

$$\boxed{I_r = \frac{U_{r\, SC}}{\sqrt{(R_r/s)^2 + X_{\sigma r\, SC}^2}}}$$

This equation represents a rotor current which is driven by the blocked rotor voltage $U_{r\, SC}$ which is of line frequency i.e. the same as the stator frequency.

The variable resistance R_r / s can be split into two components: R_r and $R_r \left(\frac{1-s}{s}\right)$

While the first component represents the copper losses of the rotor winding or bars, the second variable component is that part of the rotor which develops power when the current I_r flows through it. The corresponding rotor circuit diagram with the resistive elements separated is given in Figure 3.5 b.

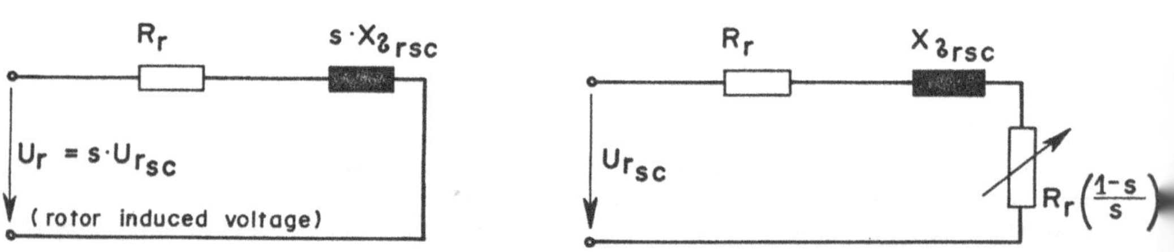

a) referred to rotor frequency

b) referred to stator frequency

FIGURE 3.5 :
Rotor circuit diagram on a per phase basis

From the diagram of Figure 3.5 we can now describe the flow of power from and to the rotor on a per phase basis (for motor operation):

Rotor (electro-magnetic) input power = rotor copper loss + rotor power developed (mechanical)

$$I_r^2 \frac{R_r}{s} \quad = \quad I_r^2\, R_r \quad + \quad I_r^2\, R_r \frac{1-s}{s}$$

Note that the power supplied to the rotor is in the form of magnetic power transmitted across the air gap between the stator and the rotor iron (see also Figure 3.7 below). To arrive at the total (electrical) input power of the machine, the losses in the stator circuit must be added.

The stator phase winding with the resistance R_s and the leakage reactance $X_{\sigma s}$, also has a magnetizing branch with the reactance X_m and a resistive element, the core loss R_{Fe}. The power furnished to the magnetizing branch is used for a large part to set up the magnetic field (aligning the magnetic axis of the ferromagnetic material to the alternating polarity) represented by the reactance X_m. It depends on the quality of the core material (magnetization curve incl. saturation) but also on the air gap. A large air gap increases the reactive power requirements to generate the magnetic field.
The other part of the power to be furnished to the magnetizing branch is an active or real-power component and is used to compensate the losses due to eddy currents and the hysteresis losses in the core (stator iron). Figure 3.6 shows the stator equivalent circuit diagram.

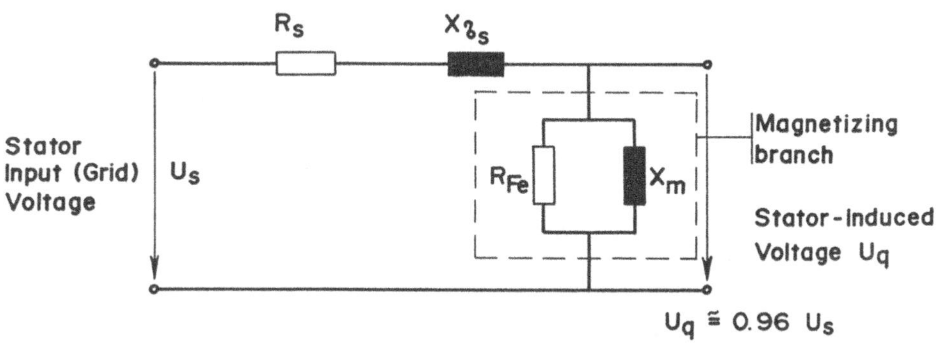

FIGURE 3.6 :
Stator equivalent circuit diagram

It now remains to combine the rotor and stator diagrams to obtain an equivalent circuit diagram for an induction motor on a per phase basis. By combining Figures 3.5 b) and Figure 3.6 we see that the two circuits are compatible if $U_{rSC} = U_q$. However, this is only the case if the turns ratio of stator and rotor windings are the same (see also transformer induced voltage, Appendix A, section 4.3). Hence, all rotor parameters are referred to stator quantities using the turns ratio.

The stator-referred quantities of the rotor circuit are then denoted R_r', $X_{\sigma r}'$, I_r' etc.

Figure 3.7 shows the complete circuit diagram of an induction machine (for one phase only). For a better understanding and allocation of the different elements of the diagram, a schematic figure of an induction machine is presented below.

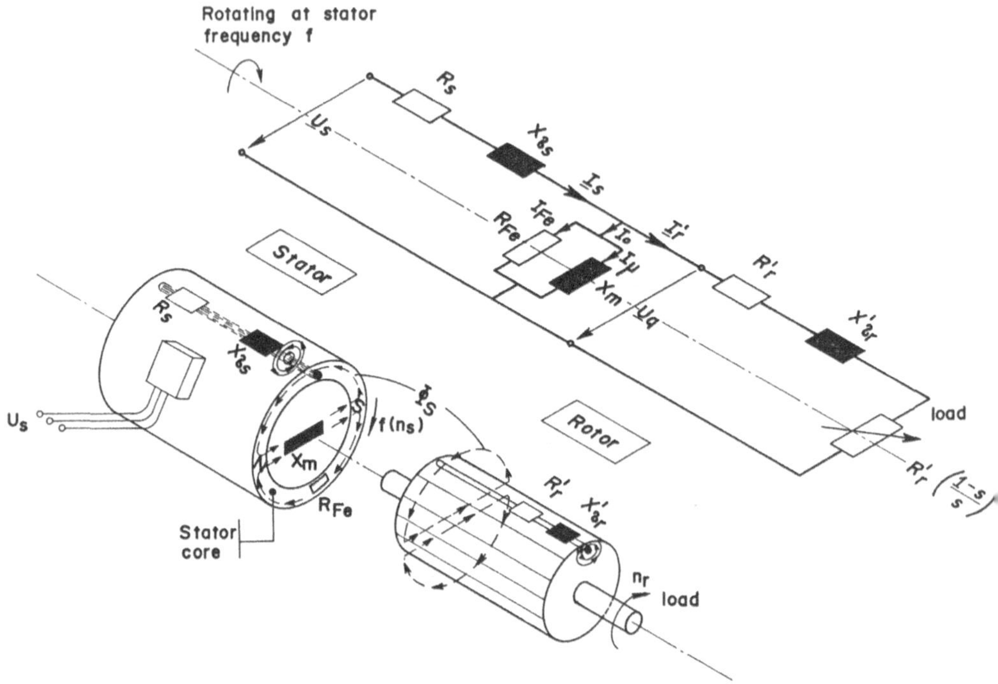

FIGURE 3.7 :
Complete equivalent circuit diagram of an induction machine

The complete equivalent circuit diagram may now be used to describe the operation of the induction machine in more details.

3.2.5 Operation at no Load

Since the motor runs at no load with a very low value of slip, there is practically no current flowing in the rotor circuit. The stator winding connected to the grid voltage will show only a low current I_o. This no-load current is mainly used to set up the magnetic field. The dominating element in the circuit is the magnetizing reactance X_m of the stator winding. Due to this large reactance as compared to the resistance of the circuit, the no-load current I_o will lag the grid voltage by nearly 90°, i.e the circuit behaves largely inductive. This explains why induction motors operate under no load and light load at a very poor power factor cos ϕ .

3.2.6 Operation under Load

We will now see how the induction motor adjusts itself automatically to an increasing load on the shaft. When a load is connected to the machine shaft (represented by the resistance $R_r'(1-s)/s$ in the equivalent circuit diagram), the rotor will be slowed down and its speed will become undersynchronous. The rotor bars will now cut the magnetic flux lines which induces voltage in the rotor winding; hence, a current I_r' will flow in the rotor circuit which develops power.
But how is this additional power drawn from the grid? The rotor current I_r' also creates flux. This flux has such a direction that it opposes the main flux from the stator current which created the rotor current and flux in the first place (Lenz's law). The resulting flux of the machine therefore tends to be reduced.

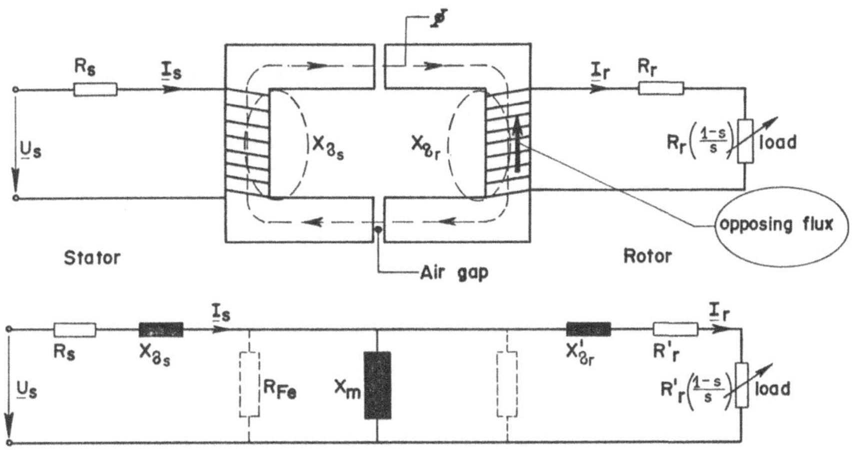

FIGURE 3.8 : Development of electromagnetic flux and stator current when loading a motor

As we know from Appendix A, current I in any coil is reduced due to the phenomenon of self-induction; i.e the magnetic field of the coil creates an induced voltage which opposes the voltage of the source. This is exactly what happens in the stator winding of the induction machine when loaded: a diminishing flux reduces the self-induced voltage of the stator winding and causes a greater difference between the grid voltage and the induced (opposing) voltage of the winding, thereby allowing more current to flow in the stator circuit. Increasing current I_s in the stator circuit brings about two things:

- the induced voltage increases to offset the tendency from the rotor circuit to reduce the flux through the core;
- the power input from the grid to the stator increases to match the increased power output of the loaded machine.

In practice, the stator winding will react almost instantaneously to increasing load and the resultant core flux remains practically constant from no load to full load.
With increasing currents, the resistive elements of the equivalent circuit diagram, especially the load Rr'(1-s)/s, rapidly become the dominating elements and the overall circuit power factor improves considerably. Figure 3.9 shows the phasor diagram of an induction motor under a) no load and b) full load with the corresponding power factors.

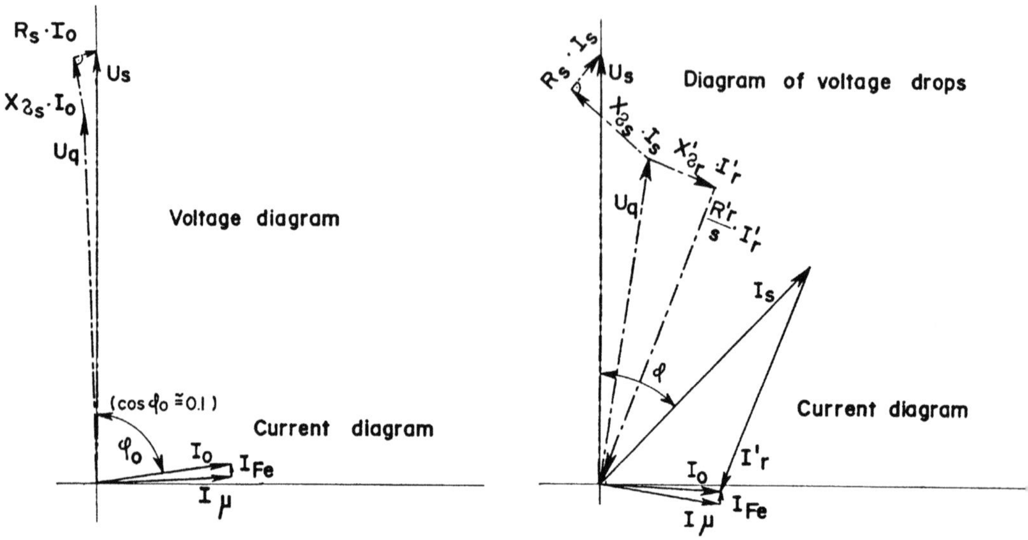

a) no load b) full load

FIGURE 3.9 :
Phasor diagram for an induction motor

3.2.7 The Flow of Power of an Induction Machine in Motor Operation

The balance of the various components of power is used to analyze the conversion of electrical to mechanical energy in an induction machine. Figure 3.10 shows the flow of power for motor operation of a three-phase induction machine.

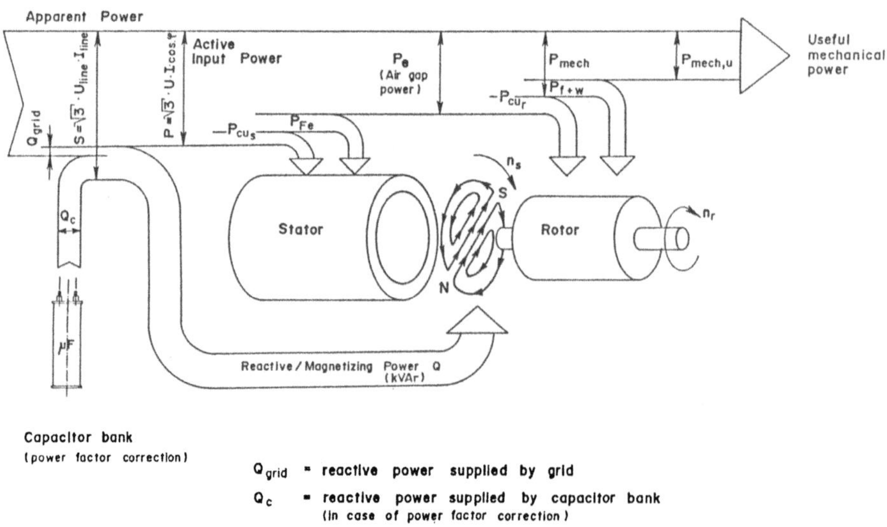

FIGURE 3.10 :
Power flow of a three-phase induction motor

The electrical power input is generally supplied by a three-phase grid. Apparent power is expressed as follows (see also Appendix A):

$$\boxed{\text{Electrical Input power (apparent): } S = \sqrt{3}\, U_{line}\, I_{line}} \qquad (3.8)$$

where U and I are the line voltage and line current of the supply network.

The reactive part $Q = S \sin\phi$ of the electrical input power is used to set up the magnetic field. Also at no load operation of the motor, it must be supplied either by the grid or (partly) by a capacitor bank (see power factor correction, section 8.6, Appendix A).

The active power absorbed by an induction machine is as follows:

$$\boxed{\text{Electrical Input Power:(active) } P = \sqrt{3}\, U_{line}\, I_{line} \cos\phi} \qquad (3.9)$$

where ϕ is the phase angle between the phase voltage and phase current of the motor.

Note that the active power P according to formula (3.8) is valid for the electrical input power of all three phases together whereas the calculations using the equivalent circuit diagram are on a per phase basis.

The useful mechanical power $P_{mech,\,u}$ on the shaft is a function of the useful torque and the rotor speed:

$$\boxed{\text{Useful Mechanical Power } P_{mech,u} = T_u\, 2\,\pi\, n_r / 60} \qquad (3.10)$$

where T_u = useful torque in Nm
n_r = rotor rotational speed in rpm

The efficiency of a machine is expressed by the ratio of output and input power. In the case of an electric motor:

$$\boxed{\text{Efficiency } \eta = \frac{P_{mech,u}}{P}} \qquad (3.11)$$

The difference of input and useful mechanical power constitutes the losses of the machine:

$$P - P_{mech,u} = \Sigma \text{ losses}$$

The following four types of losses may be distinguished:

- **Mechanical:** P_{f+w} These include all rotational losses such as friction in bearings and windage losses due to forced ventilation for cooling

- **Iron or core losses:** P_{Fe} Hysteresis and eddy current losses as a result of normal flux density changes. Iron losses in the rotor can be neglected since the flux density changes in the rotor iron are small (rotor rotates near the rotational speed of the magnetic field)
- **Copper losses** or Joule effect: The losses due to the ohmic resistance of the stator and rotor circuit
  stator: $P_{cu,s} = 3\, R_s\, I_s^2$
  rotor: $P_{cu,r} = 3\, R_r\, I_r^2$ where R = rotor or stator resistance per phase

- **Additional losses** occur in the rotor windings due to induced eddy currents; these losses are in the order of 0.5 % of rated mechanical power of the machine and can be neglected for small machines.

The power transmitted from the stator to the rotor is called air gap or **electromagnetic power Pe**. It is the electric input power P minus copper and iron losses in the stator.

$$\boxed{P_e = P - P_{Cu_s} - P_{Fe}} \qquad (3.12)$$

Since power may also be expressed by torque times angular velocity Ω, we can define an electromagnetic torque generated by the stator and acting on the rotor:

$$P_e = T_e\, \Omega_s = T_e\; 2\pi\, n_s/60$$

The electromagnetic torque is the same in both the stator and the rotor but the angular velocity of the rotor and the stator (field) are different: $\Omega_r = \Omega_s (1 - s)$

Thus the mechanical power P_{mech} developed in the rotor due to the constant electromagnetic torque T_e is lower than P_e:

$$\boxed{P_{mech} = T_e \Omega_r = T_e \Omega_s (1 - s) = P_e (1 - s)} \qquad (3.13)$$

The difference between the air gap and the mechanical power corresponds to the copper losses (Joule effect) in the rotor:

$$P_{Cu_r} = P_e - P_{mech} = s P_e = 3 R_r' I_r'^2 .$$

To obtain useful mechanical power $P_{mech,u}$, the losses due to friction and windage are to be taken into account:

$$\boxed{P_{mech,u} = P_{mech} - P_{f+w}} \qquad (3.14)$$

The useful torque on the machine shaft thus becomes:

$$\boxed{T_{mech,u} = \frac{P_{mech,u}}{\Omega_r} = \frac{P_{mech,u}}{2\pi\, n_r/60}} \qquad (3.15)$$

3.3 The Induction Machine in Generator Mode

3.3.1 General

An induction motor can be operated as a generator by driving it above its synchronous speed. The flow of power is, however, reversed: a prime mover, such as a hydraulic turbine, supplies mechanical power to the shaft of the machine. The same losses as described in section 3.2.7 above occur, but in generator operation, they must be subtracted from the mechanical input power. The same formulae and relationships as developed under 3.2.7 above apply but slip $s = (n_s - n_r)/n_s$ in oversynchronous operation becomes negative in sign as is torque. This must be taken into account in engineering computations, although in general discussions with manufacturers the negative sign may sometimes be omitted. Figure 3.11 shows the flow of power of an induction machine in generator operation.

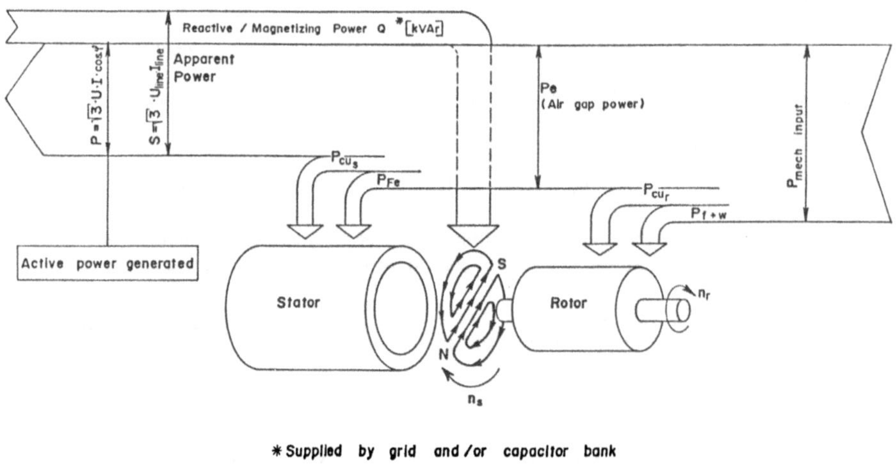

FIGURE 3.11 :
Flow of power for an induction machine in generator operation

It is interesting to see, that the flow of the reactive power Q is not reversed when the induction machine operates as a generator. The reactive power required to produce a magnetic field, can neither be generated by the prime mover (turbine) nor by the squirrel-cage rotor itself (no rotor terminals). The induction machine can therefore only operate and supply active power if connected to an external source of reactive power. When connected to a grid with other power sources, the reactive power might be supplied by synchronous generators working in the overexcited range.
In **stand-alone operation**, a battery of capacitors must be connected in parallel to the stator winding to form a resonance circuit which will supply the necessary reactive power to the induction machine. It will then work as a self-excited generator. Figure 3.12 shows a schematic circuit diagram of a self-excited asynchronous generator.

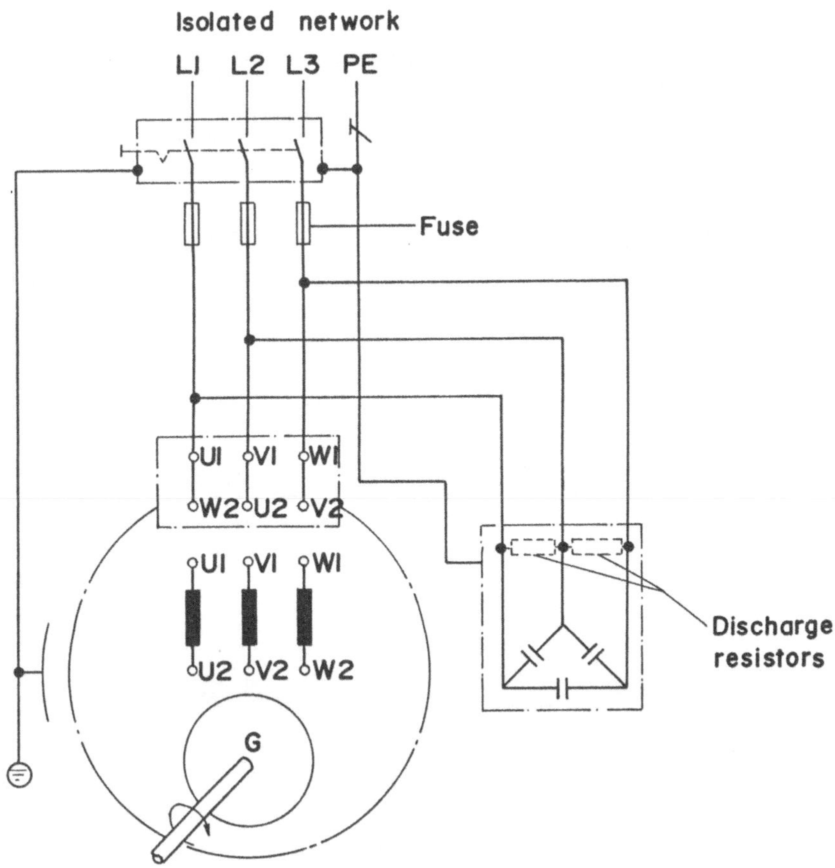

FIGURE 3.12 :
Self-excited three-phase induction generator with delta-connected capacitors

An induction motor of a given rating cannot deliver the same electrical power in generator mode as it would absorb from a grid in motor operation since the losses (copper, iron, friction and windage) reduce the output. Theoretically, the mechanical input power could be increased to make up for the losses and arrive at a higher electrical output; however, this is limited by the fact that too high an input power would soon overload the machine, i.e. overheating and finally burn the stator windings. Therefore, generator operating conditions are determined by the stator current which must not exceed the rated motor current which the windings are designed for.
Assuming, that voltage is the same for both motoring and generating, this selection condition can be expressed in power terms: Apparent (output) power in generator operation must not exceed apparent motor (input) power (see also Chapter 4, Selection).

3.3.2 Reactive Power Requirement and cosϕ in Generator Operation

When comparing the power flow of Figure 3.10 (motoring) and Figure 3.11 (generating), it may be appreciated that the air gap power P_e is higher in generating mode than in motor operation although apparent power S is the same for both conditions. This is caused by the fact that the power compensating for the stator losses (iron and copper) must be transmitted across the air gap when generating but not when motoring. Transmitting an increased power also requires more magnetizing current to set up the magnetic field which links stator and rotor. Increasing magnetic power can only take place on the expense of a reduction of the permissible active power output since apparent power is limited by the overloading condition (overheating). A shift of active to reactive power is represented by a drop in the power factor cos ϕ. It has generally been observed, that the power factor of an induction machine is at least several % lower when generating than when motoring.

This phenomenon can also be seen from the equivalent circuit (see Figure 3.13): when motoring, the induced voltage Uq is lower than the grid voltage Us because of the voltage drop across the stator circuit elements, while in generator operation, Uq must be larger by the same amount to supply nominal grid voltage at the terminals. Hence, the increase of magnetization power requirements from motor to generator operation.

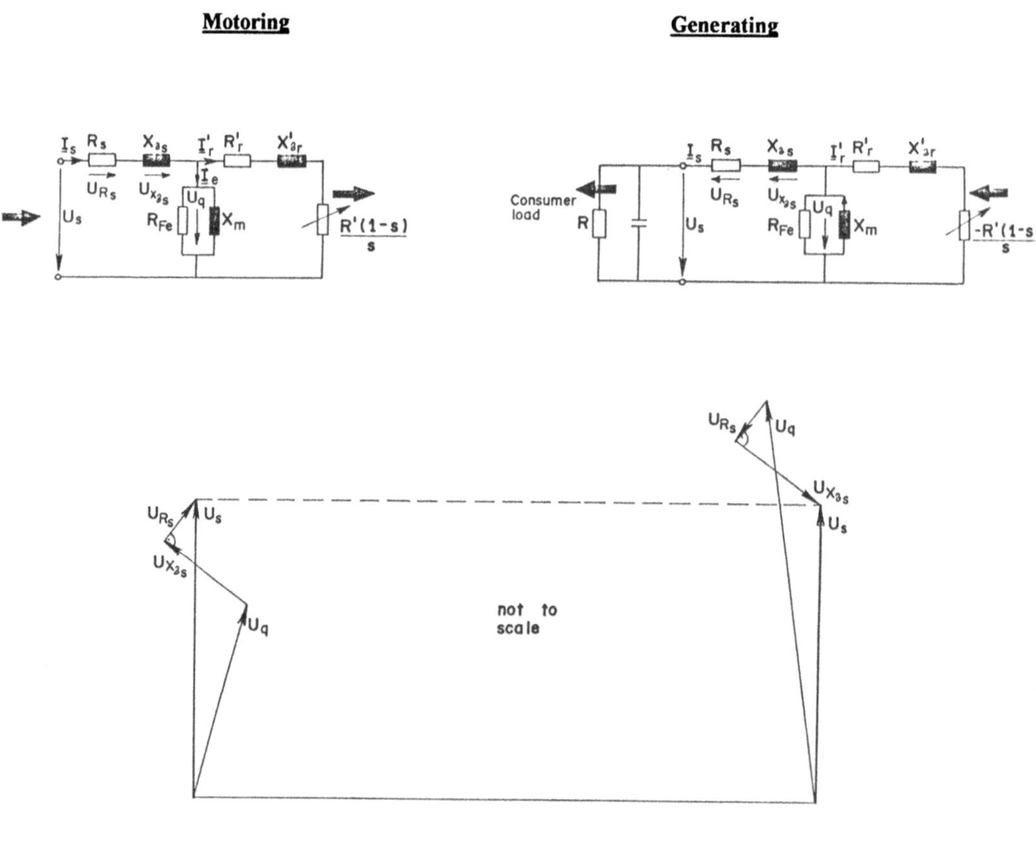

U_s = constant grid voltage

U_q = induced voltage

U_{R_s} , $U_{X_{\sigma s}}$ = voltage drop in stator elements

FIGURE 3.13 :
Equivalent circuit diagram for motor and generator operation and variation of magnetization requirements

3.3.3 *The Effect of Saturation*

A decreasing power factor in generator operation as compared to motoring becomes even more distinct due to the effect of saturation. The iron material of the stator and rotor core shows good ferromagnetic quality over a limited range, i.e. induction only increases linearly with the applied magnetization current in the coil up to a certain point (the knee) from where the curve begins to flatten. Figure 3.14 shows a typical magnetization curve of a core material.
The increase of magnetization requirements in generator operation as explained above tends to shift the operating point into the range of magnetic saturation for many modern induction motors. In order to keep the induced voltage Uq at the required level (i.e. Uq minus voltage drop across stator must still be close to nominal grid voltage, see Figure 3.13), the magnetization current and with it the reactive power demand of the induction generator must increase further. The result is a poor power factor and a corresponding loss of active power.

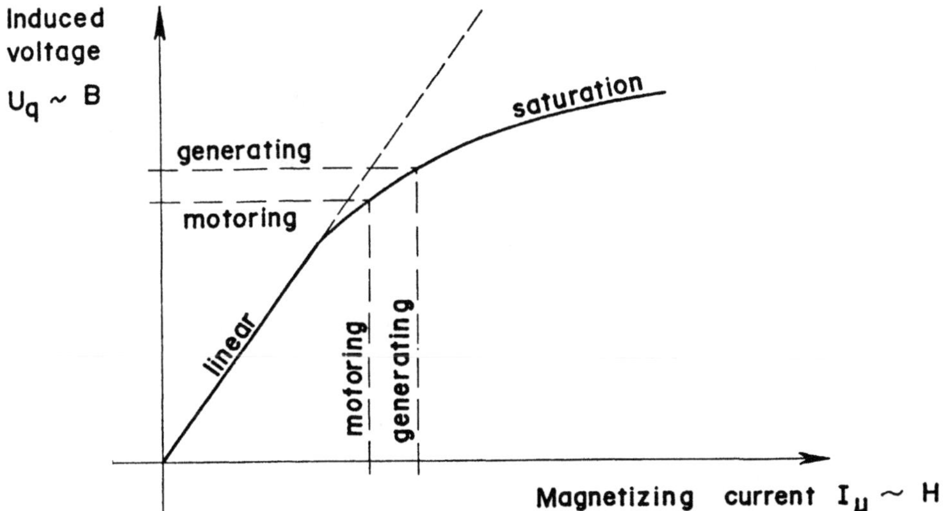

FIGURE 3.14 :
Typical magnetization curve of a common core material and increase of reactive current in generator operation

3.3.4 *The Self-Excited Generator*

As explained above, an induction machine can work as a generator in parallel to a large network which will supply the necessary reactive power to magnetize the magnetic circuits of the induction machine. When operating on an isolated grid, the magnetizing current may be furnished by capacitors (see Figure 3.12 above).
When starting the turbine / generator unit, there is no reactive current available in the stator winding to produce a magnetic field nor can the capacitors supply reactive power without being charged by inductive power first. So, how can the process be started?
Because stator and rotor iron cores have been magnetized during previous operation, they usually maintain a small amount of magnetism called residual magnetism or remanence. In Figure 3.15, this residual magnetism is represented by the offset of the magnetization curve. Driven by the turbine, the rotor of the induction machine starts to rotate and magnetic flux lines of this residual magnetism cut the stator windings upon which voltage U_1, the so-called remanent voltage, is induced.

Voltage U_1 now charges the capacitor which will supply the magnetizing current I_1 (see Figure 3.15, capacitor characteristic) back to the stator winding during the following AC half-period. I_1 in turn increases the magnetization of the machine and the corresponding voltage U_2 is produced. This process is repeated until the induced current I_i (driven by the induced voltage U_i) and the capacitor current I_c are in equilibrium. This will be the case at the intersection point P of the two curves of Figure 3.15.
This process has taken place at no-load conditions. By selecting a suitable capacitor C (μF), the no-load voltage U_0 can be set as required. At the absence of residual magnetism in the rotor, a battery (4.5 to 12 V) could be connected for a few seconds to the stator terminals (one phase only) at standstill to restore a small magnetic field in the machine which will start the process of self-excitation. A DC-voltage source which allows approximately nominal current to flow is usually sufficient .

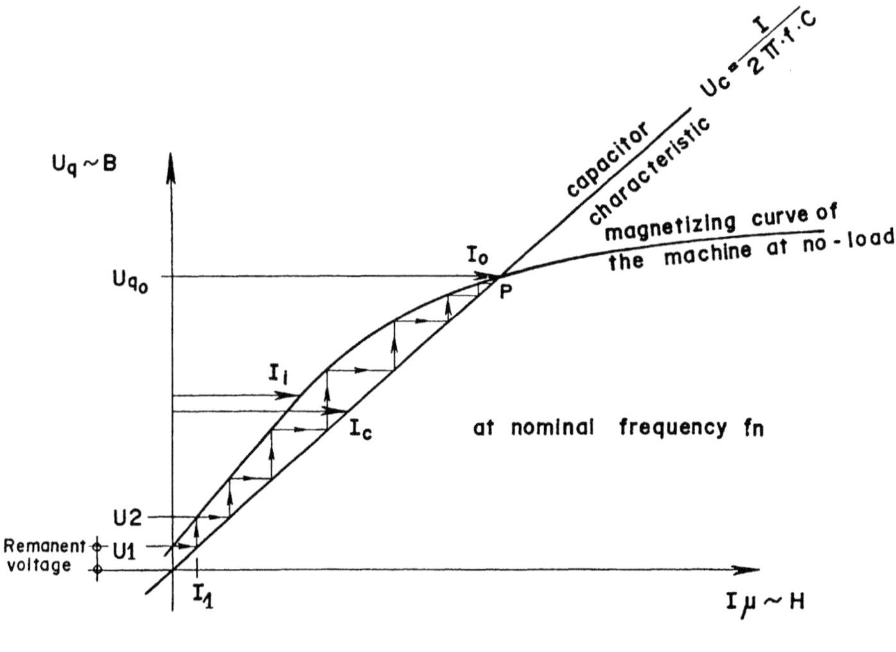

FIGURE 3.15 :
Self-excitation of an induction machine by a capacitor in parallel at no load

3.3.5 The Problem of Voltage and Frequency Control of the Induction Generator in Isolated Operation

Once the induction machine has established its no-load voltage, the consumers may be connected. As explained previously, the induction machine requires increased magnetization when it is loaded, i.e. when mechanical power is to be converted into electrical power. However, the selected capacitors can supply a constant reactive current at nominal frequency and voltage only, which is determined by the capacitor characteristic (straight line, see Figure 3.16).
Assuming that no-load voltage U_{qo} corresponds to nominal voltage U_n, the reactive current $I_{\mu o}$ is entirely used to set up the no-load magnetic field. In order to create an electromagnetic torque (which will transmit power) an additional magnetizing current ΔI_μ must be available. From Figure 3.16 we can see that this additional magnetizing current (represented by the difference between the capacitor and the no-load characteristic) is only available if the voltage at the machine terminals is allowed to drop. Maximum load is reached at the maximum value of the additional magnetizing current $\Delta I_{\mu\,max}$ corresponding to the voltage U_x . Further loading would cause an immediate breakdown of the voltage, i.e. the machine demagnetizes and does no longer deliver electrical power (Figure 3.16 a).

In fact, the maximum load, before demagnetization occurs, is further limited, since the speed of the prime mover decreases under load, so that a frequency inferior to the no-load frequency occurs. For decreasing frequency (from f_o to f_1 in Figure 3.16 b) both capacitor and no-load characteristics change: they move closer together. The corresponding maximum load (with ΔI_μ max, just before breakdown) will be lower than in case a). Figure 3.16 b) shows this behaviour.

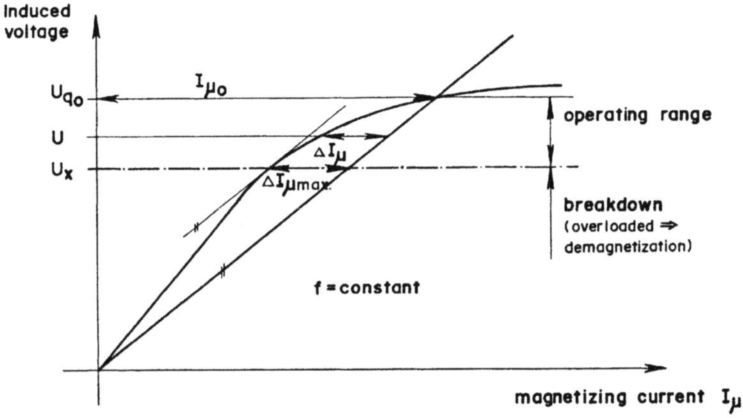

a) loading at constant frequency

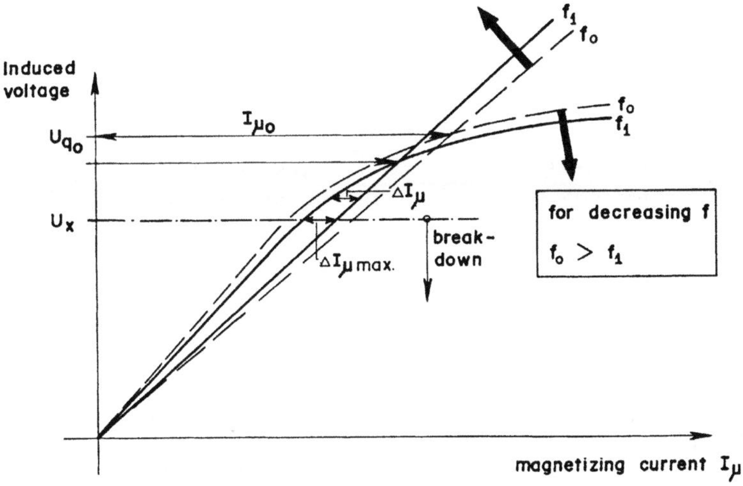

b) loading at decreasing frequency

FIGURE 3.16 :
Development of voltage at increasing load and demagnetization limit of an IMAG

It may be appreciated that the maximum value of the additional magnetizing current $\Delta I_{\mu\,max}$ is higher for core materials showing a distinct knee. Furthermore, power output of an IMAG in stand-alone operation can be increased by connecting higher rated capacitors. However, using larger capacitors will create overvoltage at no-load. Connection of the capacitors in steps, i.e. a first battery for the self-excitation process and a second battery to allow for the magnetization requirements of the load might improve the situation. Additionally, an increase of frequency can make more magnetizing current available for the load since the capacitor and no-load characteristics diverge for increasing frequency just as they shift together for decreasing frequency (see Figure 3.16 b)).

Regulation and control of voltage and frequency in relation to the electrical load and its variation will be treated in details in Chapter 5.

3.4 Induction Machine Characteristics

3.4.1 Torque

When comparing induction motors or motors in general, one of the most important parameters is the torque-speed relationship. Figure 3.17 shows a typical torque-speed characteristic of a squirrel-cage induction machine.

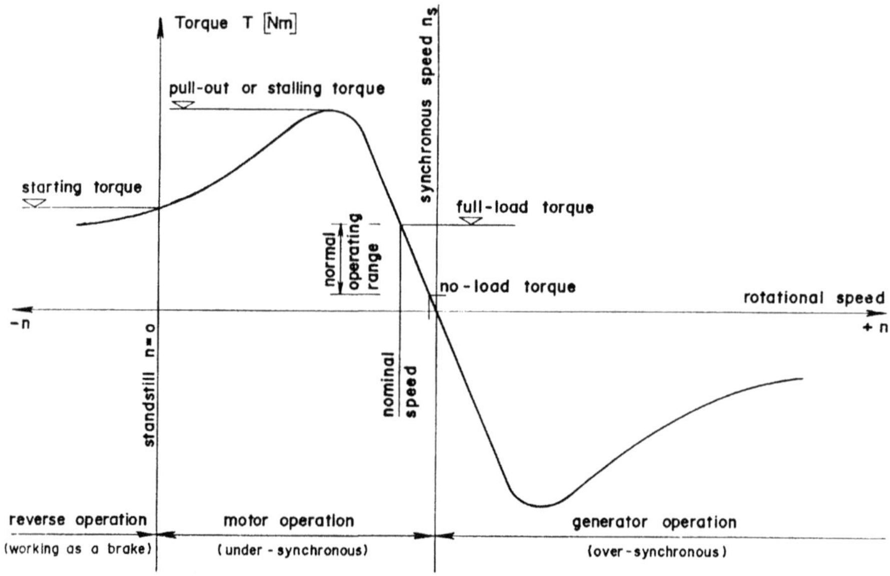

FIGURE 3.17 :
Typical torque-speed characteristic of a squirrel-cage induction machine

The torque-speed characteristic of Figure 3.17 shows a distinct peak of the torque. Loading the motor beyond that pull-out torque will cause the machine to stall and no stable operation is possible. The range between full-load torque and pull-out torque can neither be used for permanent operation because of overheating. Thus, the actual operating range both in motoring and generating represents only a small part of the complete characteristic. However, motors may have to be used to start machines under load; starting torque including the torque development up to the nominal point is therefore important. Depending on the application of the motor, the required starting torque may be high such as when driving a positive displacement pump or a crane.
Let us now consider, under which conditions a high starting torque may be produced. At normal operating speed, slip is very low and the rotating magnetic field cuts the rotor bars at a very low rate (frequency of induced voltage a few Hertz only). Inductive reactance X_r of the rotor circuit (short-circuited bars) is very small; the rotor behaves practically resistive which means that the rotor current is in phase with the induced voltage ($\cos\phi_r = 1$). Consequently, the induced rotor current will reach its maximum always at that place of the rotor periphery where the magnetic field density, i.e at the poles, is highest. Therefore, the torque produced will be maximum, too (see Figure 3.18 a).

If we look at the conditions when starting, i.e. rotor at standstill, the frequency of the rotor-induced voltage is the same as the magnetic field frequency. The inductive reactance of the rotor circuit ($X_r = 2\pi f L$) becomes important due to its being proportional to the frequency. If we assume the rotor bars being of very low resistance (large diameter), the rotor circuit will behave mostly inductive, i.e. $\cos \phi = 0$. In this case, the rotor current will lag the rotor-induced voltage by 90° . The magnetic field of this rotor current would not produce torque when superimposed on the stator field, since the resulting force couples always equal each other out (see Figure 3.18b).

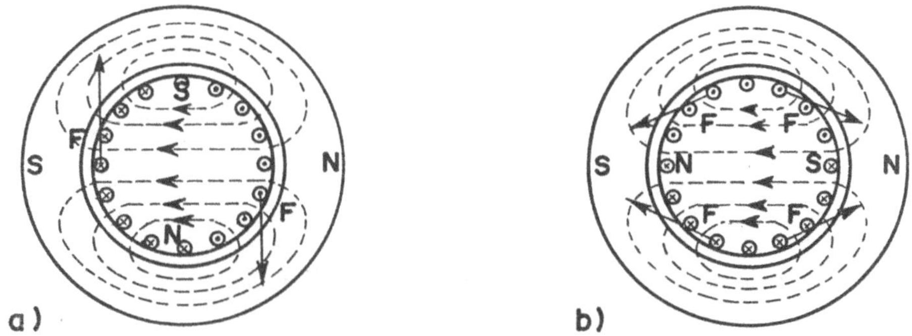

FIGURE 3.18 :

Force couples acting on the rotor of an induction motor with a) rotor $\cos \phi = 1$ and b) rotor $\cos \phi = 0$

From Figure 3.18 it becomes obvious that starting torque can only be produced if the ohmic resistance of the rotor is high as compared to its reactance, i.e. rotor $\cos\phi$ above 0. On the other hand, high efficiency under normal running conditions can only be maintained if the losses due to the rotor resistance are as low as possible; this implies high starting currents and low starting torque. Thus the rotor design is usually a compromise between the two opposing effects.

3.4.2 Rotor bar construction

Squirrel-cage motors are classified according to national standards which are not fully compatible from one country to another. Most standards, however, use the ratios of starting torque to rated torque and starting current to rated current as the main parameters. These ratios are mainly influenced by the rotor bar construction. Figure 3.19 shows four typical rotor bar designs and their torque-speed characteristics (motor operation only).

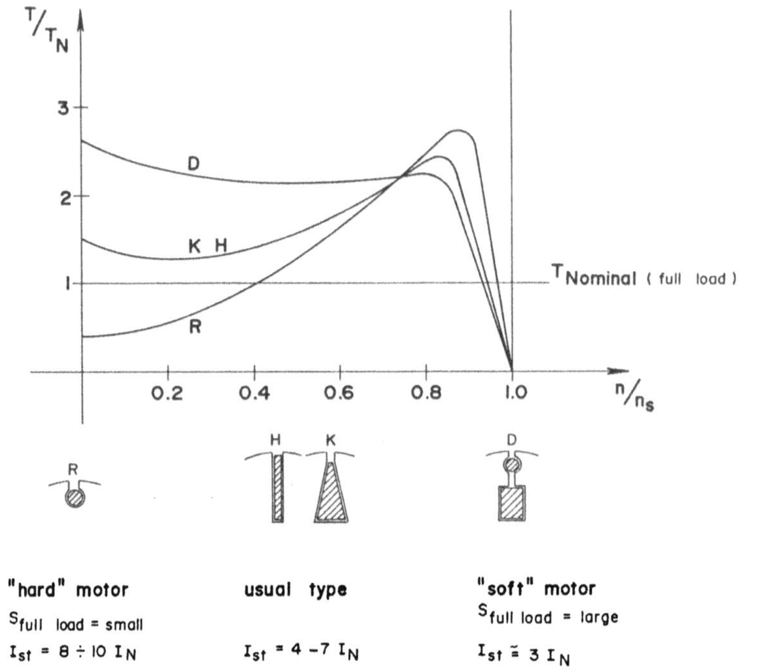

FIGURE 3.19 :
Torque-speed characteristics of squirrel-cage induction motors of various rotor bar constructions and corresponding starting current and slip ratios

As shown above, high starting torque requires a high-resistive rotor circuit; this may be achieved with circular bars of small diameter but with the expense of high copper losses and poor performance during normal operation. The double-cage and deep-bar rotors as shown in Figure 3.20 overcome this problem in an elegant way. The outer bar of small cross-sectional area placed close to the air gap is highly resistive while the inner bar with larger cross section deeply embedded into the rotor iron is of low resistance. At start-up and lower speeds, the frequency of the current in the rotor bars is high; it therefore produces high leakage reactance in the inner bar where the flux can easily develop into the rotor iron and create the opposing voltage (self-induction) which consequently suppresses the current in that cage. The outer bar on the other hand does not develop high reactance because leakage flux is limited by the air gap. The rotor current will therefore tend to flow in the outer bar and, due to its resistive nature ($\cos \phi > 0$), create good starting conditions (high torque at low current).
As the motor speed increases and consequently the rotor frequency decreases (and with it the effect of the leakage reactance), the rotor current will essentially flow in the deeper, low-resistance winding. This ensures high performance in the normal operating range.

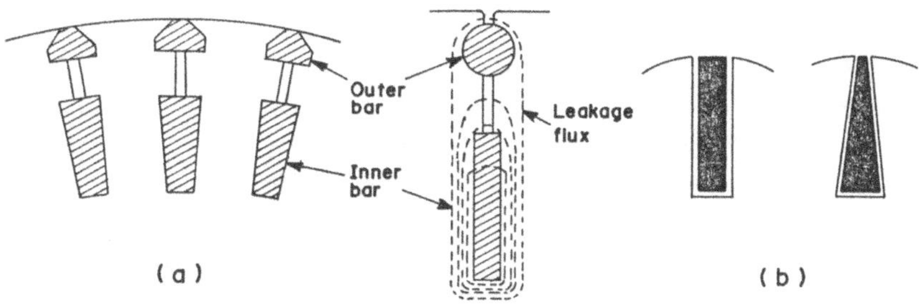

FIGURE 3.20 :
Some rotor bar constructions: a) double cage rotors and leakage flux b) deep bar rotors

Figure 3.19 above also shows that pull-out torque is essentially the same for the different types of rotor bar designs. This is due to the fact that pull-out torque is independent of the rotor winding resistance. Increasing resistance of the rotor circuit increases the slip at which maximum torque occurs, but its magnitude is almost unchanged. The linear part of the torque-speed curve is more inclined for the high resistance high starting torque motors; i.e these motors show large slip (drop of speed) when loaded and are therefore called "soft" motors while low starting torque indicates "hard" motor behaviour with little loss of rotational speed when loaded.

3.4.3 Presentation of Induction Machine Performance

Apart from the torque-speed characteristic, several other performance values of a machine are of interest. Efficiency η , power factor $\cos \phi$ at the motor terminal, motor input current I, slip s and torque T are plotted against mechanical power P_{mech}. These curves generally comprise the normal motor operating range only.

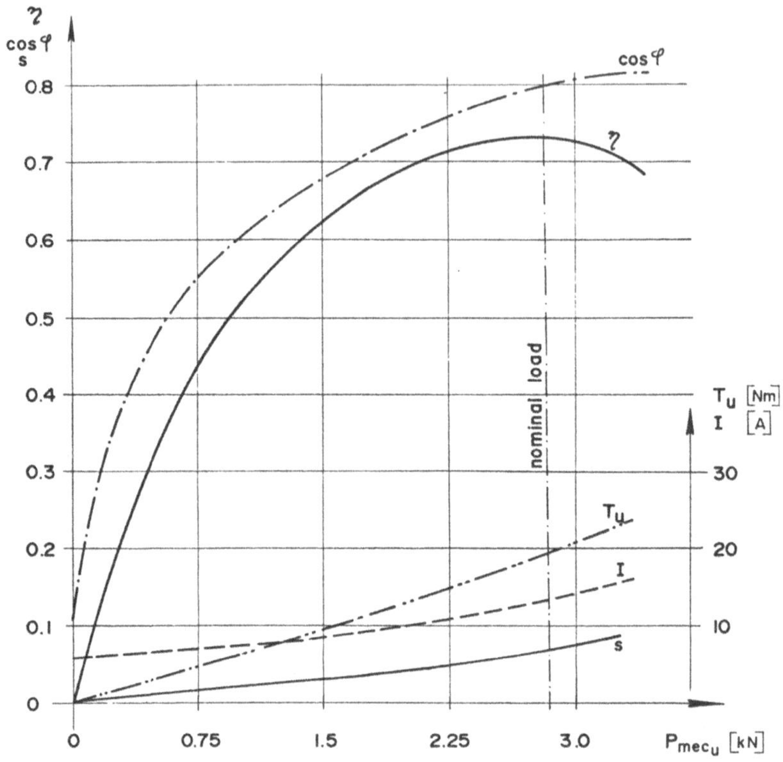

FIGURE 3.21 :
Typical characteristics of an induction motor

When the induction machine is to be used as a generator, it is more convenient to plot the performance curves against electrical power P_{el} rather than mechanical power P_{mech}. Figure 3.22 shows the complete characteristics of an induction machine of 1.1 kW mechanical motor output.

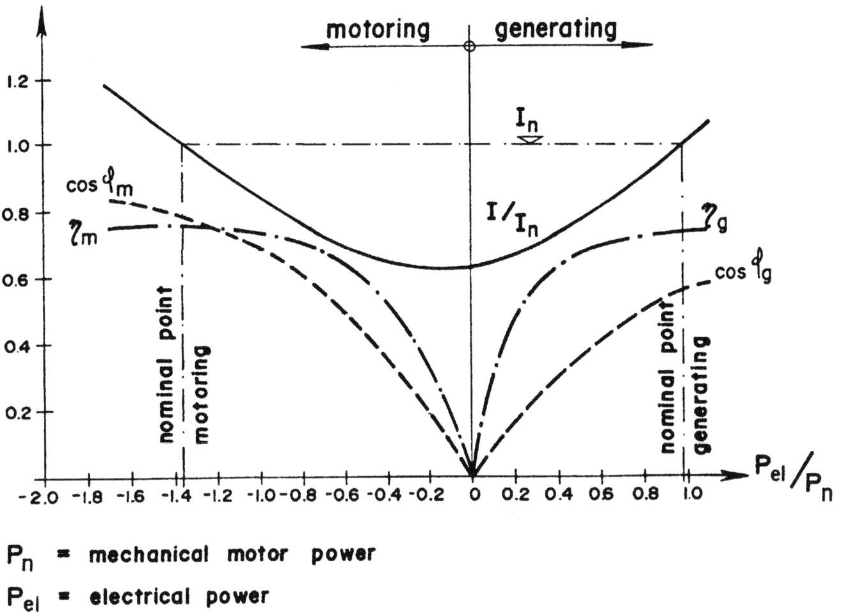

FIGURE 3.22 :
Performance curves of an induction motor for motoring and generating

3.4.4 The Circle Diagram

The circle diagram is an interesting way to present the complete performance of an induction machine for all possible operating modes; i.e motoring, generating and braking. For a constant voltage at the machine terminals and constant resistances and reactances, the electrical current varies in function of the slip and describes a circle characterizing the performance of the machine. The diagram provides for example information on the reactive (magnetizing) power demand of the machine and allows to identify where the internal losses (iron, copper, etc.) occur. The determination of the circle diagram is rather involved and requires several parameters, which can only be found by carrying out no-load and blocked-rotor tests. Additionally, the circle diagram does not take saturation effects into account and generator performance can therefore not be predicted very accurately (see Appendix C). We will use the diagram here mainly as a qualitative means to illustrate certain phenomena more clearly.

Active power is plotted on the vertical axis whereas reactive power appears on the horizontal axis (see Figure 3.23 below). At constant voltage, these distances are also a measure for the corresponding currents. Each point on the circle represents an operating point of a certain slip. Following the representation of the power triangle, the vector between zero and any point gives the apparent power (or total current) either supplied to or produced by the machine and subsequently, the angle between that vector and the vertical axis is the phase angle ϕ.
Three characteristic points on the circle can be distinguished: the no-load point P_0 (slip s = 0), standstill P_k (s = 1) and the change from motor to brake mode P_∞ (s = ∞). The straight line between P_0 and P_k is called the mechanical power line because for these two points P_{mech} = 0. The vertical distance between that line and the circumference represents the mechanical power P_{mech}. Similarly, the line between P_0 and P_∞ is the torque line because for these two points torque T = 0. Torque for any value of slip is represented by the vertical distance between that line and the circumference. Note that friction and windage losses are not yet included in the circle diagram; to obtain useful mechanical power and useful torque for motor operation, these losses must be subtracted separately. Iron losses are assumed to be constant and are represented by the distance between the horizontal axis of the diagram and the horizontal diameter of the circle.

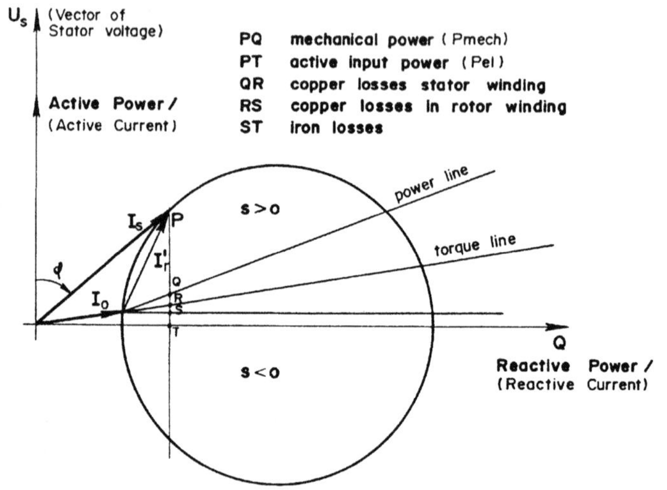

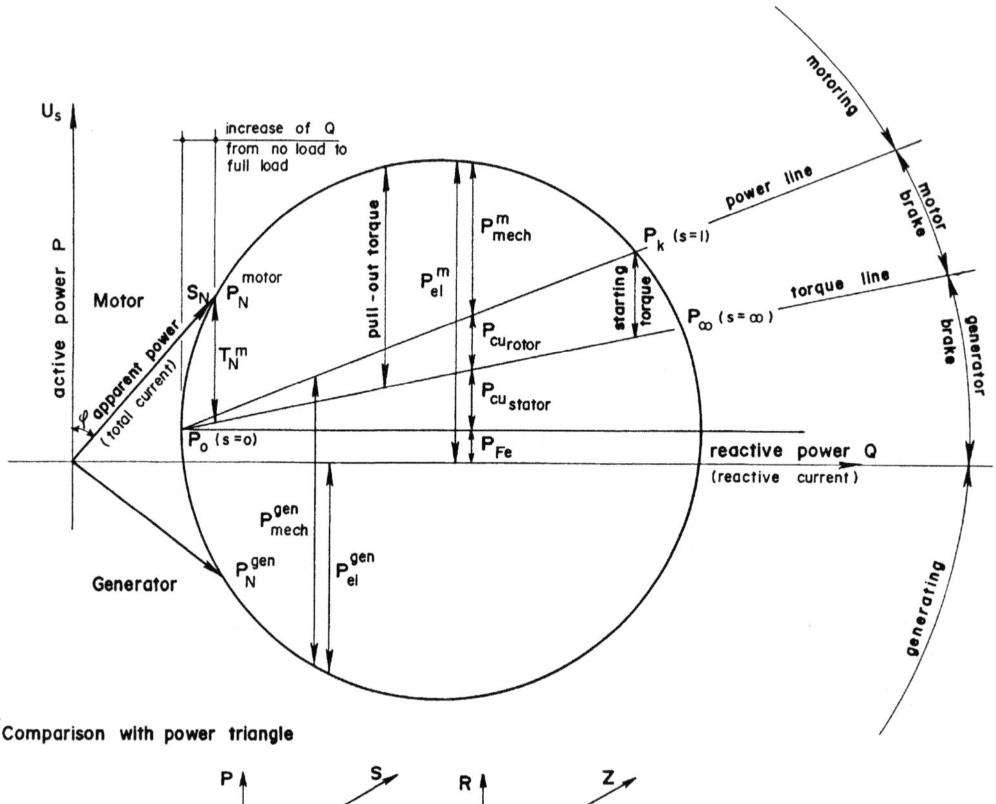

Comparison with power triangle

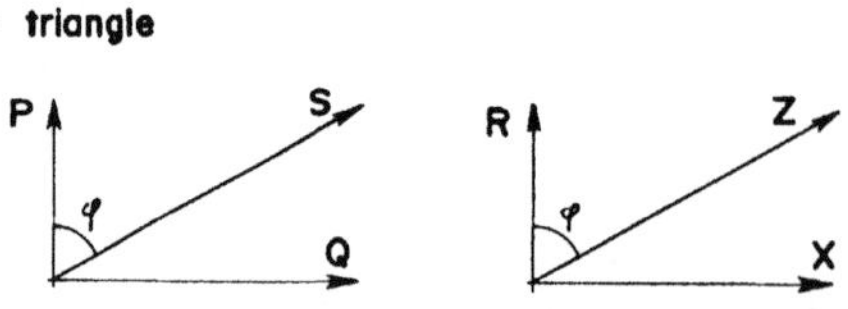

FIGURE 3.23 :
The circle diagram

4. Selecting an Induction Motor to be Used as a Generator

4.1 Mechanical Design and Rating Plate Data of Induction Motors

4.1.1 Motor Size and Standard Mounting Designs

Generally, the motor rating is designated by the mechanical power output $P_{mech,u}$ in kW and the frame size. The latter is given as the height of the shaft centre line above the frame. The same frame may be used for several power ratings. Most machines are available either for horizontal (type B) or for vertical (type V) installation and may be foot-mounted or flange-mounted. The shaft can also be designed for driving direct-coupled machines or it may receive a pulley for a belt drive. For permissible shaft loads in both axial and radial direction refer to manufacturers specifications. The following table shows the most common applications and the corresponding international code (IEC International Electrotechnical Commission No. 34-7).

IM B3	2 end shields, with feet, free shaft extension, mounting on existing base
IM B35	2 end shields, with feet, free shaft extension, mounting flange A at driving end near bearing, mounting on existing base, for flange-mounting
IM B34	2 end shields, with feet, free shaft extension, mounting flange C at driving end near bearing, mounting on existing base, for flange-mounting
IM B5	2 end shields, without feet, free shaft extension, mounting flange A at driving end near bearing, flange-mounting
IM B6	2 end shields, with feet, free shaft extension, type of construction IM B3, if required, end shields rotated through 90°, wall-mounting feet LH as seen from the driving end
IM B7	2 end shields, with feet, free shaft extension, type of construction IM B3, if required, end shields rotated through 90°, wall-mounting feet RH as seen from driving end
IM B8	2 end shields, with feet, free shaft extension, type of construction IM B3, if required, end shields rotated through 180°, ceiling-mounting
IM B9	1 end shield, without feet, free shaft extension, type of construction IM B5 or IM B14 without end shield (also without bearing) at the driving end, face mounting
IM B14	2 end shields, without feet, free shaft extension, mounting flange C at driving end near bearing, flange-mounting
IM B15	1 end shield, with feet, free shaft extension, type of construction IM B3 without end shield (also without bearing) at the driving end, mounting on existing base, face-mounting

IM V1	2 end shields, without feet, free shaft extension pointing downwards, mounting flange A at driving end near bearing, bottom-flange-mounting
IM V15	2 end shields, with feet, free shaft extension pointing downwards, flange A or C at driving end near bearing, wall-mounting
IM V3	2 end shields, without feet, free shaft extension pointing upwards mounting flange A at driving end near bearing, overhead flange-mounting
IM V36	2 end shields with feet, free shaft extension pointing upwards, mounting flange A at driving end near bearing, wall-mounting
IM V5	2 end shields, with feet, free shaft extension pointing downwards, wall-mounting
IM V6	2 end shields, with feet, free shaft extension pointing upwards, wall-mounting
IM V8	1 end shield, without feet, free shaft extension pointing downwards, type of construction IM V1 or IM V18 without end shield (and without bearing) at the driving end, face-mounting at the driving end
IM V9	1 end shield, without feet, free shaft extension pointing upwards, type of construction IM V3 or IM V19 without shield (and without bearing) at the driving end, face-mounting at the driving end
IM V18	2 end shields, without feet, free shaft extension pointing downwards, mounting flange C at driving end near bearing, flange-mounting
IM V19	2 end shields, without feet, free shaft extension pointing upwards, mounting flange C at driving end near bearing, flange-mounting

TABLE 4.1 :
Mounting design of motors and standard codes

4.1.2 Rating Plate

A number of useful data are usually given on the rating plate of the motor. Figure 4.2 shows an example of a rating plate. The theoretical background of these specifications are explained in the following. In practice, rating plates seldom contain all the specifications as indicated on our sample. Missing data may be found in manufacturers brochures.

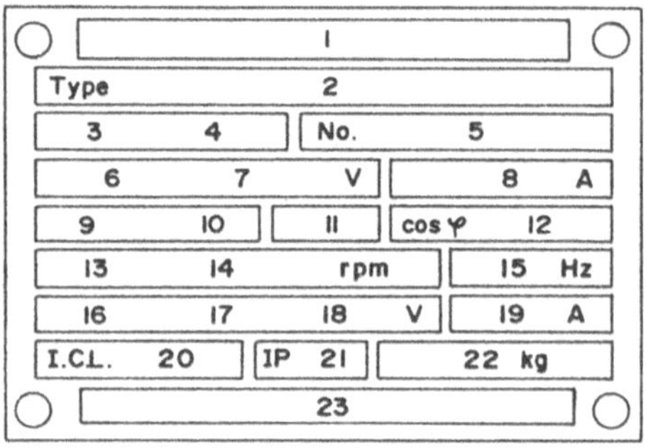

FIGURE 4.2 :
Typical rating plate of an induction motor

Item	Specification	Sample
1	**Manufacturer**	
2	**Type** of machine or model especially frame size (height of shaft centre line)	Low-Voltage **Squirrel Cage** KA7 160 M
3	**Type of current (DC, Single Phase AC or Three Phase AC)**	Three Phase **3 ~**
4	**Mode of operation**	Motor (**Mot**)
5	Product **Code / Year of manufacture**	**KA1 160M - BB 0 2 1/1991**
6	**Connection (star or delta or star/delta)**	**Y / Δ**
7	**Rated Voltage** (Line Voltage U_{line})	**380 / 220 V Y / Δ**
8	**Rated Current** (Line Current I_{line})	**23 / 40 A**
9	**Rated Power** (kVA for synchronous machines, kW for others) (see also [1]) below)	**11**
10	Unit of the power rating (kVA or kW)	**kW**
11	**Operating conditions** : (rated power usually refers to the maximum output for continuous operation = S1 or M.C.); for operation during short periods, maximum power output could be increased (no danger of overheating during a short running duration of a specified frequency); if the rating plate does not give the power for S1 or M.C., the motor must be derated for continuous operation as generator)	**S1** or M.C. (maximum continuous)
12	Power Factor for rated power (**cos φ**) (see also [4]) below)	**0.83**
13	**Direction of rotation** (can usually be either way depends on the requirements of the machine/turbine); only for unusual designs (using special fans, slip ring constructions or hydrodynamic bearings) is reverse rotation not allowed	→
14	**Rated speed**	**1450 RPM**
15	**Rated frequency**	**50 Hz**

		Example wound rotor:
16	Specifications for rotor winding (for wound-rotors only) or for excitation in the case of synchronous machines (applies to items 17 to 19)	Rotor
17	Connection of the rotor winding (applies for wound-rotor motors with slip rings only) is usually in star Y	Y
18	(Line) voltage at rotor open circuit (for wound-rotor designs only)	195 V
19	Current at slip rings for operation at rated conditions	45 A
20	**Insulation class or temperature rise** (see 2) below)	**F, 80° C**
21	**Enclosure** (see 3) below)	**IP 54**
22	**Weight**	116 kg
23	Other Information: - Moment of Inertia - Cooling - Specification of the standards applied	 0.0589 kgm2 **Fan Ventilated** IEC 34-1 (1969)

1) Derating

If the machine is operated (as motor or as generator) at an air (cooling) temperature in excess of 40° C or above an altitude of 1000 m, the rated output must be reduced to avoid overheating.

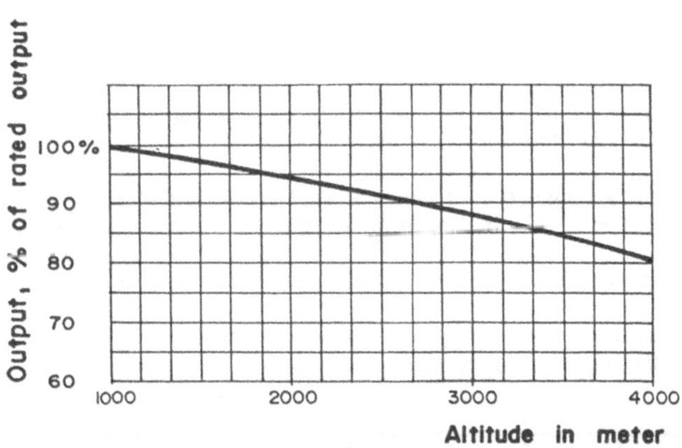

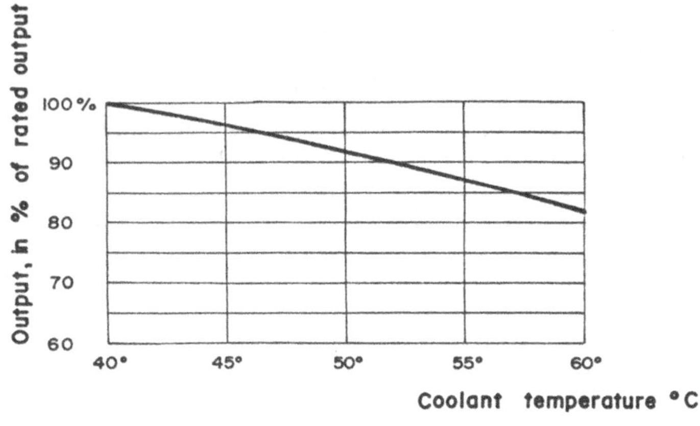

Where both air temperature and altitude are exceeded, the corresponding reduction factors should be multiplied to obtain the final reduction factor.

The altitude limit may be exceeded without reducing the power output if the air temperature does not exceed the following values:

Altitude in meter	1000	2000	3000	4000
Max. cooling air temperature ° C (Class B insulation)	40	32	24	16

2) Insulation Class and temperature rise:
The permissible temperature rise of induction motors is limited by the heat resistance of the insulation material used for the windings. The most commonly used classes for induction motors are classes B and F which comprise varnished wires with insulating materials such as glass fibre, asbestos, mica, synthetic resins and polythene materials (EPOXY etc.).

Referring to an ambient temperature of 40° C the permissible temperature <u>rise</u> of these insulation classes are:

- Class B: 80° C (maximum temperature 130° C)
- Class F: 100° C (maximum temperature 155° C)

For extreme environmental conditions (aggressive gases, etc) special insulation is required; for extremely high humidity and/or temperature differences, a tropical-proof insulation is available from some manufacturers.

3) Enclosure
The design of the motor frame is not only determined by the size of the motor but also by the protection required by the application of the machine. The extent of protection includes two effects which are standardized and designated as IP (International Protection) classes:

- protection of people from getting in contact with rotating parts or electrical components under voltage and protection against penetration of solids into the motor; defined by the first figure (0 to 6 after the sign IP)

- protection against penetration of water into the motor; specified by the second figure (0 to 8) after IP

The most commonly manufactured motors are the enclosure classes IP 23 , IP 44, IP 54 and IP 55. The figures stand for the following extent of protection:

First Figure	**Protection against solid particles**
0	no protection
2	protection against solid particles larger than 12 mm
4	protection against solid particles larger than 1 mm
5	protection against deposition of dust inside the motor; penetration of dust is not completely excluded but the amount of dust penetrating has no effect on the proper operation of the machine
6	protection against penetration of dust

Second Figure	Protection against penetrating water
0	no protection
3	protection against splash water up to an angle of 60^0 from the vertical 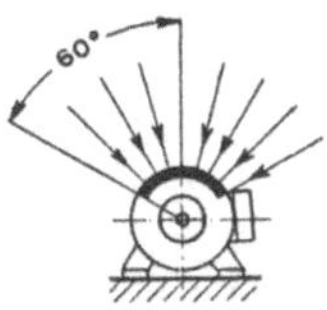
4	protection against splash water at any angle 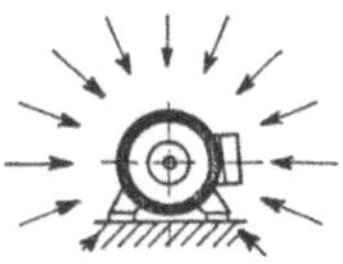
5	protection against a water jet at any angle 

4)
If the power factor is not indicated on the rating plate, it can be calculated:
For a three-phase machine:
Apparent power $S = \sqrt{3}\, U_{line}\, I_{line}$ (total input power);

The active input power $P = P_{mech,u} / \eta$

where $P_{mech,u}$ = rated power output on the shaft and η = overall efficiency. Since the power factor is defined as $\cos\phi = P / S$, we obtain:

$$\cos\phi = \frac{P_{mech,u}}{\eta \sqrt{3}\, U_{line}\, I_{line}} \qquad (4.1)$$

If both efficiency and power factor are not given, we may assume that the values of efficiency and power factors are in the same order of magnitude and we obtain:

$$\cos\phi \text{ and } \eta \approx \sqrt{\frac{P_{mech,u}}{\sqrt{3}\, U_{line}\, I_{line}}} \qquad (4.2)$$

Compare also with the general values given in Appendix B3.

4.2 General Considerations on Selection

Ideally, the selection of an IMAG should be as easy as the selection of a generator: manufacturers brochures are reviewed in search of the machine which matches the available power of the prime mover (turbine) with maximum efficiency. However, this is not the case at present due to the general lack of recorded experience with IMAGs in practical applications. The prediction of the machine performance in generator mode can neither be found in manufacturers' handbooks nor elsewhere. General rules for the selection of an IMAG may work well for certain designs whereas they will fail for others. This is especially true for the small power range we are dealing with in this manual (0.5 to 50 kW).

Induction motors of different designs may have similar output when motoring but may yield a considerably different performance when generating. This can best be illustrated with the circle diagram.

Consider two motors, A and B, of similar power output (P_n = 2.2 kW); Motor B, however, has been designed to start high inertia loads, i.e. its starting torque is higher than that for motor A. This special feature of motor B has been attained by a more elaborate rotor bar construction (double cage, see section 3.4.2 above) but only at the expense of a slightly lower efficiency and a lower power factor cosϕ at the nominal operating point. This is represented in the circle diagram (see Figure 4.3a below) by a shift of the circle towards higher currents (or power) both active and reactive. Note, that the mechanical power output remains essentially the same for both motors. Comparing the magnetization curves of these two machines (Figure 4.3b), we can see that the double-cage motor operates deeper in the saturated range (higher magnetization current at nominal voltage), hence its higher reactive power demand. It is also the effect of saturation which causes the circle diagram to appear somewhat different from theory: the circular shape of the curve is not maintained when changing from motoring to generating; a distinct knee appears at about no-load current (slip = 0). This feature is even more distinct for the highly saturated machine.

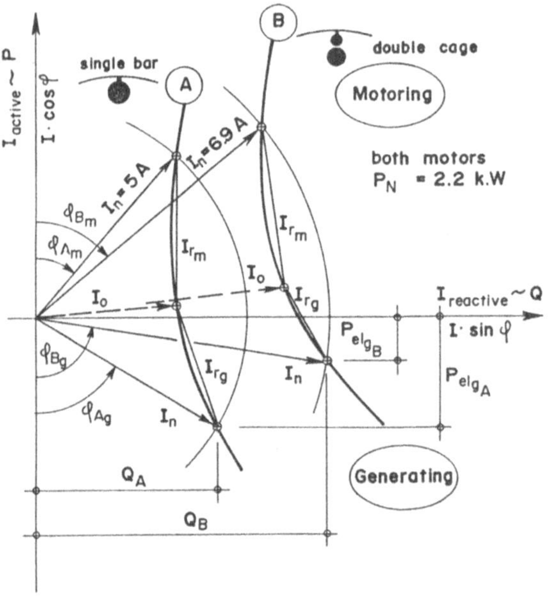

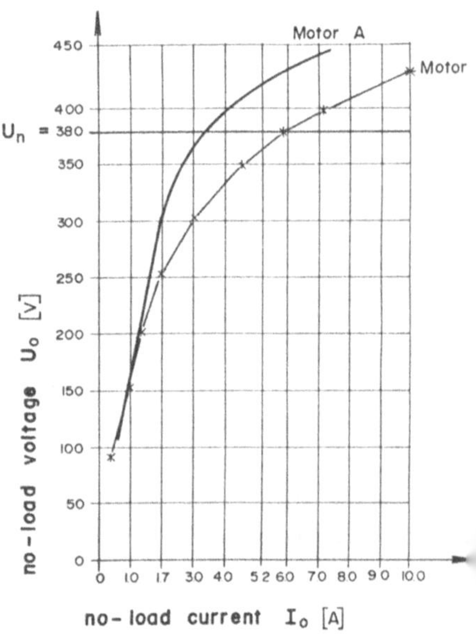

FIGURE 4.3 :
a) Circle diagram for two motors of similar motor power but of different designs
b) Magnetization curves of the two motors

In generator operation, the nominal stator current I_s (or apparent power S) must not exceed the rated current in motor operation in order to avoid overheating. Therefore, the maximum operating point in generating mode can be found at the same distance I_s from zero but on the lower half of the circle. Figure 4.3 shows that the electrical power output when generating is lower than one would expect from theory (ideal circle diagram): the knee in the curve considerably reduces the output. In the case of the highly saturated motor B, the electrical output is not even half of that of motor A. Furthermore, the reactive power requirement of IMAG B has increased more from motor to generator operation than that of motor A. When operating as a stand-alone electricity generating unit, the required capacitors to supply the reactive current will add additional costs to IMAG B. Hence, the essential rule for the selection of an IMAG can be summarized as follows:

Saturated motors are unsuitable as IMAGs and should be avoided

The problem is, how to distinguish the highly saturated machines from the less saturated ones which will most probably yield better performance in generator mode. Since testing machines before installing them in an MHP is not a feasible solution for the practical engineer, we have to define a number of parameters which are easily accessible, (e.g. from rating plates or manufacturer's brochures) and which will allow to distinguish suitable from unsuitable IMAGs.

4.3 Practical Selection Methods

4.3.1 Selection Criteria - Suitable Machines

Section 4.2 has indicated that the rotor bar construction is probably the main parameter for an accurate selection of an IMAG. Unfortunately, manufacturers seldom indicate the rotor bar design of their machines on generally available data sheets. Therefore, we have to identify the machine through criteria which are in one way or other related to the rotor bar design.

- *Efficiency*: When power flow is reversed from motor to generator operation, it is the losses and their distribution which may have unpredictable effects on the generator performance. High efficiency motors have by definition relatively small losses which, when generating, cannot cause surprising results. Therefore, **high efficiency** is the key criterion and engineers should always aim at high efficiency machines despite the possibly higher costs of such motors. Appendix B3 provides general values of efficiencies in function of rated motor power.

- *Full-load slip or rated speed:* As stated earlier, "hard" motors, i.e. those which do not deviate much from synchronous speed when loaded, usually show good performance as generators. When comparing machines from various manufacturers, select those with **lowest full-load slip**. General indications of low slip values (speed plotted against power) distinguishing "hard" from "soft" motors are given in Appendix B4.

- *Ratio between starting torque and pull-out torque:* A **low starting torque T_{st} or a low ratio of starting torque to pull-out torque T_{max}** characterizes motors with low rotor resistance. This indicates a rotor bar design without special arrangements (special designs such as double-cage rotors are detrimental to generator performance). For the small power range (0.5 kW < P_n < 50 kW) the following ratios should be observed:

  $$T_{st} / T_n < 2.0 \quad \text{and} \quad T_{st} / T_{max} < 0.9$$

  where T_{st} = starting torque
  T_n = nominal torque
  T_{max} = pull-out or stalling torque (these ratios are usually given in manufacturer's brochures)

- *Starting current:* The desired low starting torque for IMAGs usually coincides with high starting currents which may cause difficulties when motoring. Since starting with load is not relevant for IMAGs, high starting currents should not be regarded as a disadvantage; quite on the contrary: **high motor starting currents** indicate suitable IMAGs. The following ratios may be taken as approximate reference values:

  $$\text{for small motors } P_n < 2.0 \text{ kW: } I_{st} / I_n > 5$$

  $$\text{for } P_n > 3.0 \text{ kW: } I_{st} / I_n > 7$$

- ***Power Factor*** $\cos\phi$: The power factor indicates the quality of the magnetic circuits. However, when selecting an IMAG, it might be misleading: modern induction motors use a relatively high level of saturation to achieve good performance in motor operation; when generating, the effect of saturation increases and performance drops (see above). The power factor is therefore not a suitable criterion to select an IMAG.

- ***Saturation:*** The negative effects of saturation can be reduced if machines with a higher voltage rating than necessary for the application are selected. Manufacturers in the UK for example manufacture also standard ***induction motors with 415V / 50 Hz rating*** (in other countries only available for an extra charge). If these machines are used as generators for a 380 V or 400 V distribution, saturation problems are unlikely to occur since magnetization current will not increase excessively when the application changes from motoring to generating (see magnetization curve of Figure 4.3b above and section 3.3.3 for the theoretical background).

- ***Internal connection (Y or Δ):*** Three-phase standard motors are available with star or delta connected windings. Depending on the application as IMAG, the following internal connections should be ordered:

  - **for parallel operation**: star or delta connection may be used but Δ is to be preferred to reduce problems with third-harmonic currents; refer to regulations of the electricity board

  - **for stand-alone three-phase operation without transformer**: delta-connected IMAGs can ensure that third-harmonic currents do not cause problems for consumers. The disadvantage is that only one voltage level is available due to the omission of the neutral (see also Chapter 6: Safety and Protection)

  - **for stand-alone three-phase operation with transformer**: delta or star connection of the IMAG is possible but the neutral has to be isolated. The transformer must be delta-connected in the primary and star-connected in the secondary to provide a neutral point

  - **for stand-alone single-phase distribution** using a 3-phase IMAG with C-2C capacitors (see also Chapter 6): star or delta connection is possible

- ***Enclosure:*** Mass-produced induction motors are usually manufactured in three standard enclosure types, namely: IP 23, IP 44, IP 54/55. The latter, ***IP54 or 55 is the preferred form of mechanical protection for an IMAG.*** Its totally enclosed design is not only advantageous in view of the sometimes rough conditions at a micro-hydropower plant (always wet; dusty if the prime mover drives machinery such as saw mills, flour mills, hullers, etc. in combination or alternating with the IMAG) but also in view of the efficiency of the machine: Efficiencies of the IP 23 design are usually lower than those of the IP 54/55 type because the open construction of the former evacuates more heat and therefore operates at increased losses. The advantage of the IP 23 lies in its slightly lower weight and in its lower price. The IP23 type induction motor is available from about 10 kW rated motor power upwards.

- ***Rated synchronous speed (number of poles):*** As a general rule, the number of pole pairs should be kept as low as possible in order to save on costs, weight and complexity of machines. Additionally, efficiencies and power factors of machines decrease as the number of poles increases.

  When using the machine as a generator in a hydropower plant, a further parameter has to be considered: the overspeed capability of the machine. It becomes important because of the possible runaway of the turbine. Runaway speed of turbines may reach twice or more the nominal speed which would mean 6000 rpm or more at runaway for the two-pole IMAG! Standard motors can seldom stand these high rotational speeds. We therefore recommend that motors of four or more poles (corresponding to 1500 rpm or lower at 50Hz) be selected. These models usually have standard rotors and bearings which are used for all motor types (including speed modulated machines with up to 200 % nominal speed) throughout. Hence, these IMAGs will be able to withstand turbine runaway without modifications.

  Direct-coupling of machines to the turbines is to be preferred wherever possible (see also Manual on Pumps Used as Turbines; J-M. Chapallaz/GATE 1992)

4.3.2 Selection Procedure

From the findings of the previous sections we can now state the following selection conditions:

a) electrical limit ($I < I_n$)

The electrical output power (= apparent power S) when generating should not exceed the electrical input power when motoring so that the windings are used to their full capacity but without being overloaded ($I < I_n$).

Motoring: mechanical output $P_n = S * \cos\phi_m * \eta_m$ where

apparent power $S = \sqrt{3}\, U_{line}\, I_{line}$

Generating: electrical output $P_{el_g} = S * \cos\phi_g$

Condition: apparent power $S_{motoring} = S_{generating}$

$$P_{el_g}\,max = \frac{\cos\phi_g}{\eta_m \cos\phi_m} P_n \qquad (4.3)$$

Note: Generator power factor $\cos\phi_g$ should not be mistaken for the power factor of the load. $\cos\phi_g$ is exclusively determined by the IMAG constants and the operating slip.

In Formula 4.3 above, only the power factor in generator mode ($\cos\phi_g$) is unknown; motor efficiency and power factor are usually given by manufacturers brochures (mean values see Appendix B). Based on test results of various sizes of standard motors, a graph indicating approximate ratios of $\sin\phi_g / \sin\phi_m$ (versus rated motor power P_n) has been plotted (see Figure 4.4 below). Using actual motor data ($\cos\phi_m$; η_m and P_n), the maximum electrical output Pel_g can be computed using formula 4.3 above.

Note that Figure 4.4 is only valid for standard non-saturated motors; machines which are likely to operate at a high saturation level when generating must first be eliminated from the selection charts. The criteria given in section 4.3.1 above may be applied.

If only saturated motors are available, their use as generators can still be a feasible solution. However, the operating conditions (voltage and frequency or internal connection of windings) must be altered to arrive at a reasonable electrical output; for more details see Section 5.5).

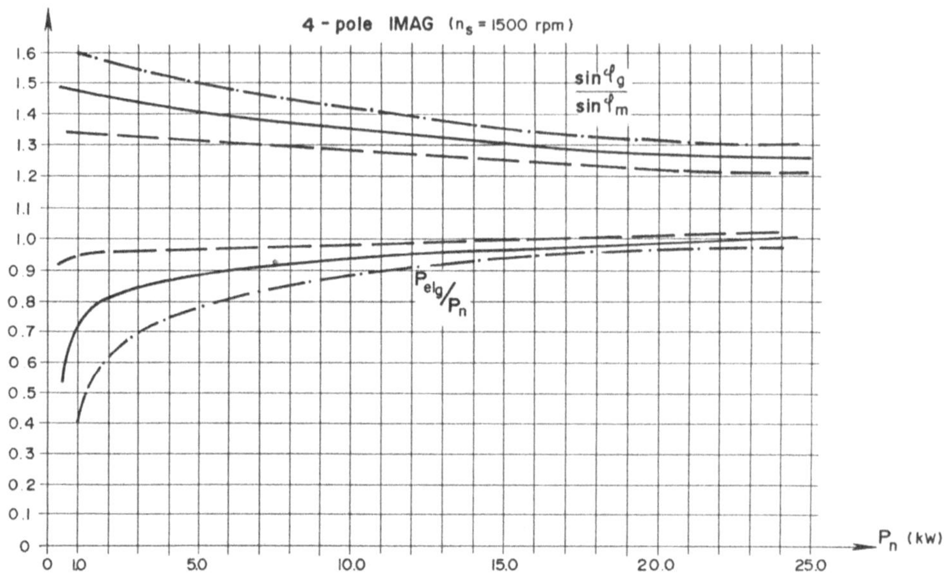

FIGURE 4.4 :
Ratio of generator sin ϕ to motor sin ϕ in function of rated motor power P_n (based on test results)

The product $\eta_m * \cos\phi_m$ in Formula 4.3 takes a fairly constant value for each size of standard motor (see Appendix B3). We can therefore directly indicate an approximate ratio of generator output to rated motor power. Such a graph is plotted in Figure 4.4 using the mean value of the sinϕ - ratio (full line) and the upper and lower limit of the scatter (broken lines); it may be used as a preliminary selection criteria in the absence of detailed motor data.

b) thermal limit (permissible temperature rise)

In order that the machine does not burn, a thermal limit must be defined. The temperature rise generated by the losses of the machine must not exceed the permissible temperature rise of windings. Most losses increase with increasing load on the IMAG. A relatively conservative assumption is the following:

the permissible temperature rise when generating is the same when motoring or in other words, the thermal limit of the machine is attained if the losses when generating are equal to the losses when motoring

This is a conservative assumption since the increased cooling effect for fan-cooled motors operating in oversynchronous mode (rotational speed above n_s = generating) is not taken into account.

The thermal limit will not necessarily produce the same operating point as the electrical limit defined by the stator current ($I < I_n$); the latter deals with the condition in the stator only while a number of additional effects/losses influence the thermal equilibrium of the machine:

- **stator copper losses** (increasing with the square of the stator current):

$$P_{loss_s} = 3 * R_s * I_s^2 \qquad \text{where } R_s = \text{stator resistance and } I_s = \text{stator current.}$$

If the electrical limit of $I = I_n$ is fulfilled, the thermal conditions of the stator will essentially be the same when changing from motoring to generating

- **rotor copper losses** (increasing with the square of the rotor current):

$$P_{loss_r} = R_r * I_r^2 \quad \text{where } R_r = \text{rotor resistance and } I_r = \text{rotor current.}$$

From Figure 4.3 above (circle diagram) we can see that the rotor current tends to be reduced when changing from motoring to generating. This is especially true for saturated machines (with a distinct knee in the circle diagram).

- **iron losses** (increasing approximately proportional to the magnetic power Q). Since reactive power increases when changing from motoring to generating, we might expect higher magnetic losses in generator mode especially for saturated machines.

- **friction losses** in bearings remain approximately constant

- **windage losses** (friction in air gap) increase with the third power of the rotational speed:

$$P_{loss_W} \sim n^3$$

For the power range $P_n < 20$ kW windage losses increase between 20 to 30 % when changing from motoring (undersynchronous speed) to generating (oversynchronous speed).

Test results revealed that the effects increasing and those decreasing the losses equal each other out in the small power range we are dealing with in this manual. The generator operating point determined by the permissible temperature rise (thermal limit) is therefore always close to the point of nominal current. This is shown in Figure 4.5 where the development of currents and the losses for both motor and generator operation have been plotted (test of a 2.2 kW induction motor, non-saturated).

In the course of the selection procedure of IMAGs, we will therefore assume that the thermal limit corresponds to the electrical limit of induction machines in generator mode.

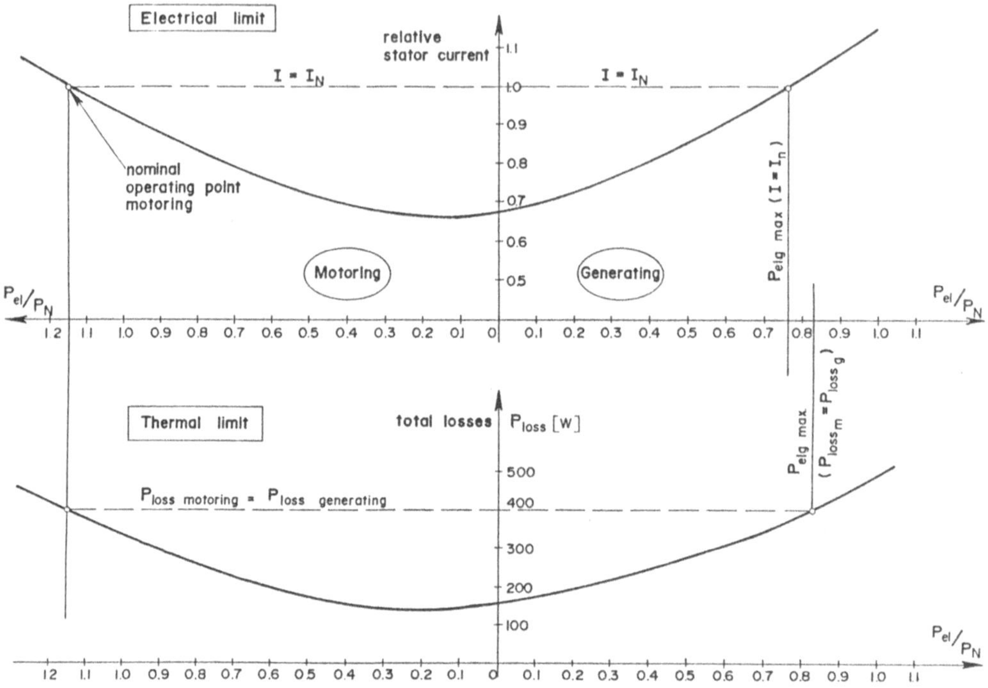

FIGURE 4.5 :
Comparison between the electrical limit ($I < I_n$) and the thermal limit for generator operation

Assuming that total losses when motoring are equal to the losses in generator mode we can now determine the necessary input power from the prime mover to drive the IMAG:
Condition:

$$P_{loss\ motoring} = P_n\,(1/\eta_m - 1) = P_{loss\ generating}\ max$$

$$P_{mec_g}\ max = P_{el_g} + P_{loss} = P_{el_g} + P_n\,(1/\eta_m - 1)$$

and the corresponding generator efficiency (at maximum electric power output according to Formula 4.3 above):

$$\eta_{g\ max} = \frac{1}{\frac{P_n}{P_{el_g}}(1/\eta_m - 1) + 1} \qquad (4.4)$$

The formulae above contain only values which have either been determined in the course of this section (Pel_g from Formula 4.3) or which are known from motor data. The permissible mechanical input power of the IMAG (design limit) or its corresponding efficiency can therefore be calculated without requiring further information.
In practical work, we usually know the turbine output, which is determined by the head and flow conditions of the site and we have to find the IMAG which matches this turbine output as closely as possible. For a pre-

selection of the motor size required, Figure 4.6 below may be used. Note that the generator input power is equal to the turbine output minus possible transmission losses of belts or gear boxes; thus:
$P_{mec_g} = P_{mec_t} * \eta_{trans}$. Approximate values of transmission efficiencies are given in Appendix B6.

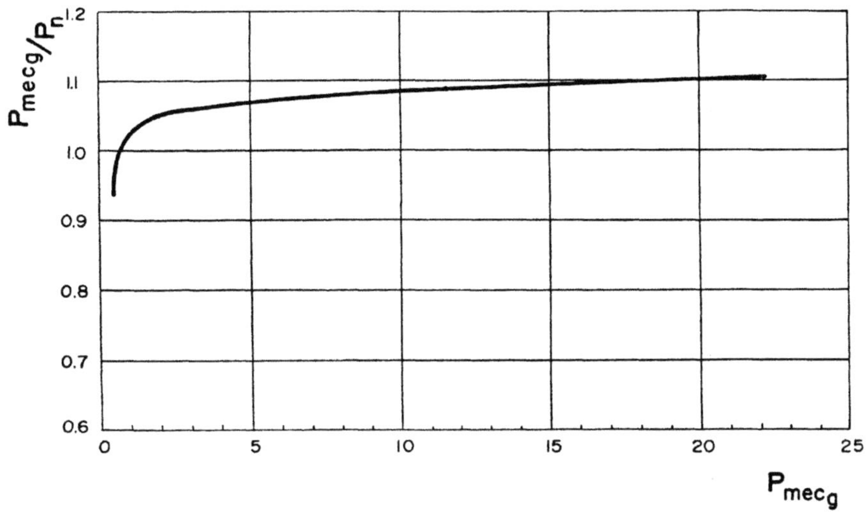

FIGURE 4.6 :
Pre-selection of the motor size (rated motor power) required for a specific application as IMAG (using standard motor efficiencies and mean value of sin ϕ-ratio)

4.3.3 Generator efficiency and power factor at part-load

Due to the inevitable inaccuracies of the selection procedure, the selected IMAG should be slightly **overrated**; i.e. the calculated maximum electrical output P_{el_g} max should be about 20 to 30 % above the layout point of the plant (P effective). With this safety margin the IMAG will not overheat even if the actual turbine and generator operating point turns out to be different from the calculated design value.
Overrating the IMAG has the disadvantage that both efficiency and power factor will be lower if the machine operates below its maximum output. We assume that the drop of generator efficiency and power factor will be in the same order of magnitude as the drop of the corresponding motor values. Table 4.8 below provides general values of efficiencies and power factors for part-load operation.

It has been found that similarity between motor and generator efficiency (and power factor) curves only exists if the point of maximum generator output is referred to the motor part-load value of P_{el_g} max / P_n (and not motor full-load 4/4). The generator curves for both efficiency and power factor are then plotted from this point downwards assuming direct proportionality with the motor curves:

$$\eta_g = \frac{\eta_m}{\eta_{m\,o}} \; \eta_{g\,max} \quad \text{and} \quad \cos\phi_g = \frac{\cos\phi_m}{\cos\phi_{m\,o}} \; \cos\phi_{g\,max}$$

This may not be very scientific but correlation with test results was found to be good. Figure 4.7a shows this relationship for the efficiency .
In order to determine the effective operating point (corresponding to the effective turbine output), we require a graph of generator efficiency versus mechanical input power. This can be plotted using Figure 4.7a and the relationship:
$P_{mec_g} = P_{el_g} / \eta_g$ applied on a number of points (e.g. 1/4 , 2/4 , 3/4 and 4/4 of P_{el_g})

Figure 4.7 b shows such a graph. Entering with the effective mechanical input power into the graph, we obtain the effective efficiency , power factor and the corresponding effective electrical output of the IMAG. The same graph can be used to determine the output of the IMAG at turbine part flow (dry-season output).

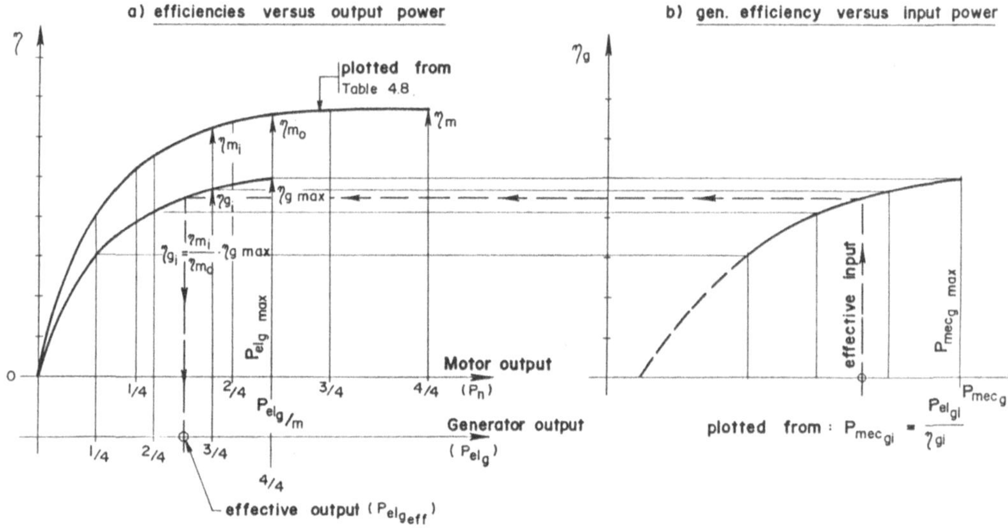

FIGURE 4.7 :
Approximation of generator efficiency curves and estimation of effective output of the IMAG

TABLE 4.8 :
Decrease of efficiency and power factor for standard motors at part-load operation (for IP54/55)

Motor efficiency at part load;

frame size < 132

η_m at part load of:

5/4	4/4	3/4	2/4	1/4
84.5	86	86	84	78
83.5	85	85	83	77
82	84	84	82	76
81	83	83	81	75
80	82	82	80	74
79	81	81	79	73
78	80	80	78	72
77	79	79	77	70
76	78	78	76	68
75	77	77	75	66
74	76	76	74	64
73	75	75	72	62
71	74	74	71	60
70	73	73	70	59
69	72	72	69	58
68	71	71	68	56
67	70	70	67	55
66	69	69	66	54
65	68	68	65	53
64	67	67	64	52
63	66	66	63	50
62	65	65	62	48
61	64	64	61	47
60	63	63	60	46
59	62	62	58	45
58	61	61	56	43
57	60	60	55	42
56	59	59	54	40
55	58	58	53	38

frame size > 160

η_m at part load of:

5/4	4/4	3/4	2/4	1/4
96	96	96	94.5	90
95	95	95	93.5	88
94	94	93.5	92	86
93	93	93	91	85
92	92	92	90	84
91	91	91	89	82
90	90	90	87	80
89	89	89	86	79
88	88	88	85.5	78.5
86	87	87	85	78
85	86	86	84.5	77.5
84	85	85	84	77
83	84	84	83	76
82	83	83	81	74
81	82	82	80.5	73
79	81	81	80	72
78	80	80	79	70
77	79	79	78	69.5
76	78	78	76	69
75	77	77	75	68
74	76	76	74	67
73	75	75	73	66
72	74	74	72	64
71	73	73	71	63
70	72	72	69	61
69	71	71	68	59
68	70	70	67	58
67	69	69	66	57

Motor power factor at part load

frame size < 132

$\cos\phi_m$ at part load of:

5/4	4/4	3/4	2/4	1/4
0.92	0.91	0.86	0.78	0.58
0.91	0.90	0.85	0.77	0.57
0.90	0.89	0.84	0.75	0.55
0.90	0.88	0.82	0.72	0.50
0.89	0.87	0.81	0.70	0.48
0.88	0.86	0.80	0.69	0.47
0.87	0.85	0.79	0.68	0.46
0.86	0.84	0.78	0.67	0.45
0.86	0.83	0.77	0.66	0.44
0.85	0.82	0.75	0.64	0.43
0.84	0.81	0.74	0.62	0.42
0.84	0.80	0.73	0.60	0.40
0.83	0.79	0.71	0.58	0.39
0.82	0.78	0.70	0.57	0.38
0.81	0.77	0.69	0.56	0.37
0.80	0.76	0.68	0.55	0.36
0.79	0.75	0.67	0.54	0.36
0.78	0.74	0.66	0.53	0.35
0.77	0.73	0.65	0.52	0.35
0.76	0.72	0.65	0.51	0.35
0.76	0.71	0.64	0.50	0.35
0.75	0.70	0.62	0.49	0.35
0.74	0.69	0.60	0.48	0.35
0.73	0.68	0.59	0.46	0.34
0.72	0.67	0.57	0.45	0.34
0.71	0.66	0.56	0.44	0.34
0.70	0.65	0.55	0.44	0.34
0.68	0.64	0.54	0.42	0.33
0.67	0.63	0.53	0.41	0.32
0.66	0.62	0.53	0.41	0.32
0.65	0.61	0.52	0.40	0.32
0.64	0.60	0.52	0.40	0.30
0.63	0.59	0.51	0.40	0.30
0.62	0.58	0.50	0.39	0.30
0.60	0.57	0.48	0.39	0.30
0.59	0.56	0.47	0.38	0.29
0.58	0.55	0.46	0.37	0.28

frame size > 160

$\cos\phi_m$ at part load of:

5/4	4/4	3/4	2/4	1/4
0.94	0.94	0.92	0.88	0.74
0.94	0.93	0.92	0.87	0.68
0.92	0.92	0.89	0.84	0.65
0.91	0.91	0.88	0.82	0.64
0.90	0.90	0.87	0.80	0.63
0.89	0.89	0.86	0.79	0.60
0.88	0.88	0.85	0.78	0.58
0.88	0.87	0.84	0.77	0.57
0.87	0.86	0.83	0.75	0.55
0.86	0.85	0.82	0.73	0.53
0.84	0.84	0.81	0.72	0.51
0.85	0.83	0.80	0.70	0.49
0.85	0.82	0.78	0.67	0.47
0.83	0.81	0.76	0.66	0.45
0.82	0.80	0.75	0.65	0.43
0.82	0.79	0.73	0.63	0.42
0.79	0.78	0.73	0.60	0.41
0.78	0.77	0.72	0.59	0.40
0.78	0.76	0.70	0.58	0.38
0.77	0.75	0.69	0.56	0.36
0.76	0.74	0.67	0.54	0.36
0.75	0.73	0.66	0.52	0.35
0.74	0.72	0.65	0.51	0.34
0.73	0.71	0.64	0.50	0.34
0.72	0.70	0.63	0.48	0.33
0.71	0.69	0.62	0.47	0.33
0.70	0.68	0.61	0.45	0.32

4.3.4 Summary of the selection procedure
(see also Chapter 8: Worked Examples)

1. Determine the turbine output power $P_{mec\,t}$ and the rotational speed of the turbine n_t for your specific application (design values of head and flow). For preliminary projects or feasibility studies Diagram B7 in Appendix B may be used to determine the type of turbine required and its corresponding rotational speed.

2. On the basis of this speed, decide whether a transmission gearing is to be used or not; It should be remembered that (for reasons related to the overspeed capability of standard motors) the maximum synchronous speed (n_s max) for IMAGs is 1500 rpm corresponding to a 4-pole machine (see section 4.3.1 above); however, the number of poles should still be as few as possible since efficiencies and power factors drop with increasing number of poles.
General indications of transmission efficiencies are given in Appendix B6.

P_{mec_g} effective $= \eta_{trans} * P_{mec\,t}$

3. Using Figure 4.6 above, an approximate size of IMAG (rated motor power P_n) which matches the mechanical input power P_{mec_g} may be determined.

4. Review manufacturers brochures in search of motors with this power rating. Decide whether the machines available are likely to be saturated or not; use the parameters given in section 4.3 above (efficiency, slip, ratio of starting torque to pull-out torque, etc. and compare with the reference values given above)

5. Calculate the maximum electrical power output of the IMAG using the motor data according to the manufacturer (P_n, η_m and $\cos\phi_m$) and an approximate generator $\cos\phi_g$ calculated from Figure 4.4 above:

$$P_{el_g}\ \max = \frac{\cos\phi_g}{\eta_m \cos\phi_m} P_n$$

Note that the nominal motor power P_n must be reduced (derated) if the machine will operate in an air (cooling) temperature in excess of 40 °C and/or at an altitude of >1000 m (see section 4.1.2 for derating factors).

6. Calculate the corresponding mechanical input power P_{mec_g} max:

$$P_{mec_g} = P_{el_g} + P_n (1/\eta_m - 1)$$

and compare with the effective input as determined by the site conditions (point 2 above)

7. a) if P_{mec_g} effective > $Pmec_g$ max, select a motor with a higher rating P_n
b) if P_{mec_g} effective < P_{mec_g}, the selected machine may be a feasible option; however, it must be verified whether a minimum safety margin is maintained, i.e. the IMAG should be slightly overrated to avoid overheating in any case. As a general rule, maximum electrical power P_{el_g} max should be about 20 to 30 % above the effective electrical output P_{el_g}eff corresponding to the layout point of the plant.
Plot the generator efficiency curve using the method proposed in section 4.3.3 above and determine graphically the effective generator efficiency $\eta_g$eff (see Figure 4.7 b above); then calculate:

P_{el_g} effective $= \eta_g$ eff $* P_{mec_g}$ effective

8. Determine the required capacitors and the protection and control equipment according to Chapter 5 below.

9. Justify the economic advantage of the IMAG as compared to a synchronous generator or a purpose-wound induction machine through an approximate financial analysis (see Chapter 7).

4.3.5 Example

A small hydropower plant of 11 kW turbine power output should be equipped with an IMAG. Using the selection procedure of chapter 4, a motor with the following specifications has been chosen:

Type: 160 L4 (four poles)
- nominal motor power $P_{mec_m} = 15\ kW = P_n$
- nominal speed $n_n = 1455$ rpm
- efficiency $\eta = 89\ \%$
- power factor $\cos\phi_m = 0.86$
- nominal current $I_{m_n} = 30$ A

The plant will operate in parallel to the national grid with 50 Hz frequency and 380 V nominal voltage.

Using the motor data and the diagrams of Section 4.3 above, we can estimate the permissible input and the corresponding output of this machine in generator mode (see also section 4.2.3 above):
From Figure 4.2 and $P_n = 15$ kW:

$\sin\phi_g / \sin\phi_m = 1.30$

with $\cos\phi_m = 0.86 \Rightarrow \sin\phi_m = 0.510$, the generator power factor becomes:

$\sin\phi_g = 1.30 * 0.510 = 0.663 \Rightarrow \cos\phi_g = 0.748$

$$P_{el_g}\max = \frac{\cos\phi_g}{\eta_m \cos\phi_m} P_n = \frac{0.748}{0.89 * 0.86} 15 = 14.66\ kW$$

$$P_{mec_g}\max = P_{el_g}\max + P_n (1/\eta_m - 1) = 14.66 + 15\,(1/0.89 - 1) = 16.5\ kW$$

$$\eta_g \max = P_{el_g}\max / P_{mec_g}\max = 14.66 / 16.5 = 0.88$$

We can see that this generator will have the proposed safety margin (overrating) since the maximum turbine power is only 11 kW while the IMAG selected could theoretically absorb 16.5 kW; generator efficiency at lower than maximum power can be estimated from motor part-load efficiency curves (see also section 4.3.3 above).

The same procedure is also used for the generator power factor at part-load operation. Table 4.8 above provides **part-load power factors** for various motor sizes. After plotting the $\cos\phi$ - curve for motoring, we draw a similar curve for generating beginning at the point $P_{el_g}\max / P_n = 0.98$ with our $\cos\phi_g \max = 0.748$ as calculated above.

Efficiency and power factor versus output power for both motor and generator operation ***generator efficiency versus mechanical input power***

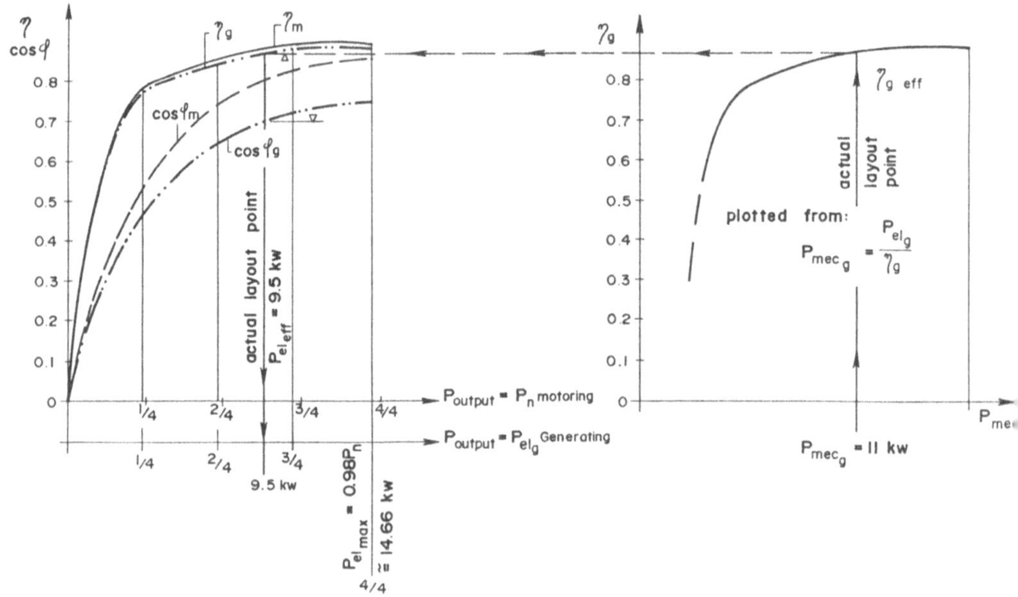

From the plot on the right, we obtain at P_{mec_g} eff = 11 kW an IMAG efficiency of η_g eff = 86.5 % and the effective electrical output becomes:

$$P_{el_g}\,eff = \eta_g\,eff * P_{mec_g}\,eff = 0.865 * 11.0\ kW = 9.5\ kW$$

which corresponds to 66 % of P_{el_g} max.
The determination of the excitation capacitance required will be shown in Section 5.1.2.

4.4 Determination of IMAG performance from test results

Theoretically, it is possible to calculate the performance of an induction machine given the resistance and the reactance values of the equivalent circuit diagram (see section 3.2.4, Figure 3.7). These parameter values can be determined by performing relatively simple tests:

- measurement of the ohmic resistance of the stator windings R_s

- no-load test: the motor is run with the load uncoupled and it therefore rotates at a very low value of slip; since almost no current is induced in the rotor, the rotor circuit can be assumed as being open. Without the influence of the rotor, the following parameters can be determined: friction and windage losses, the magnetizing reactance X_m, core resistance R_{Fe} and the stator leakage reactance X_{σ_s}.

- blocked rotor test: this test corresponds to the conditions at start-up. Since the currents at start-up are normally five to seven times rated current, this test is carried out at a greatly reduced voltage so that only nominal current flows. At this reduced voltage the magnetizing reactance X_m becomes very large and the magnetizing branch (see equivalent circuit, Figure 3.6) can be neglected. This allows to determine the resistance R_r' and the leakage reactance X_{σ_r}' of the rotor.

Having determined all values of the equivalent circle diagram, the performance of the induction machine can be calculated or plotted using the circle diagram. It has been found, however, that this method does not take into account the effects of saturation, neither in motor nor in generator mode. As can be seen from

Figure 4.3 above, saturation brings about a distinct knee in the circle which does not appear in the theoretical approach using the parameter values of the equivalent circuit diagram. Figure 4.9 shows the complete performance of an induction machine both from actual full-scale measurements and from the theoretical approach using the two points P_o (no-load) and P_k (blocked rotor) to draw the circle diagram. It can be seen that correlation of the two curves is not sufficiently accurate.

An improvement may be brought about if a third point, the nominal operating point P_n of the motor (from name plate data) is used to plot the theoretical circle. With this method, correlation with the real performance may be in the order of ± 10 to 30 %.

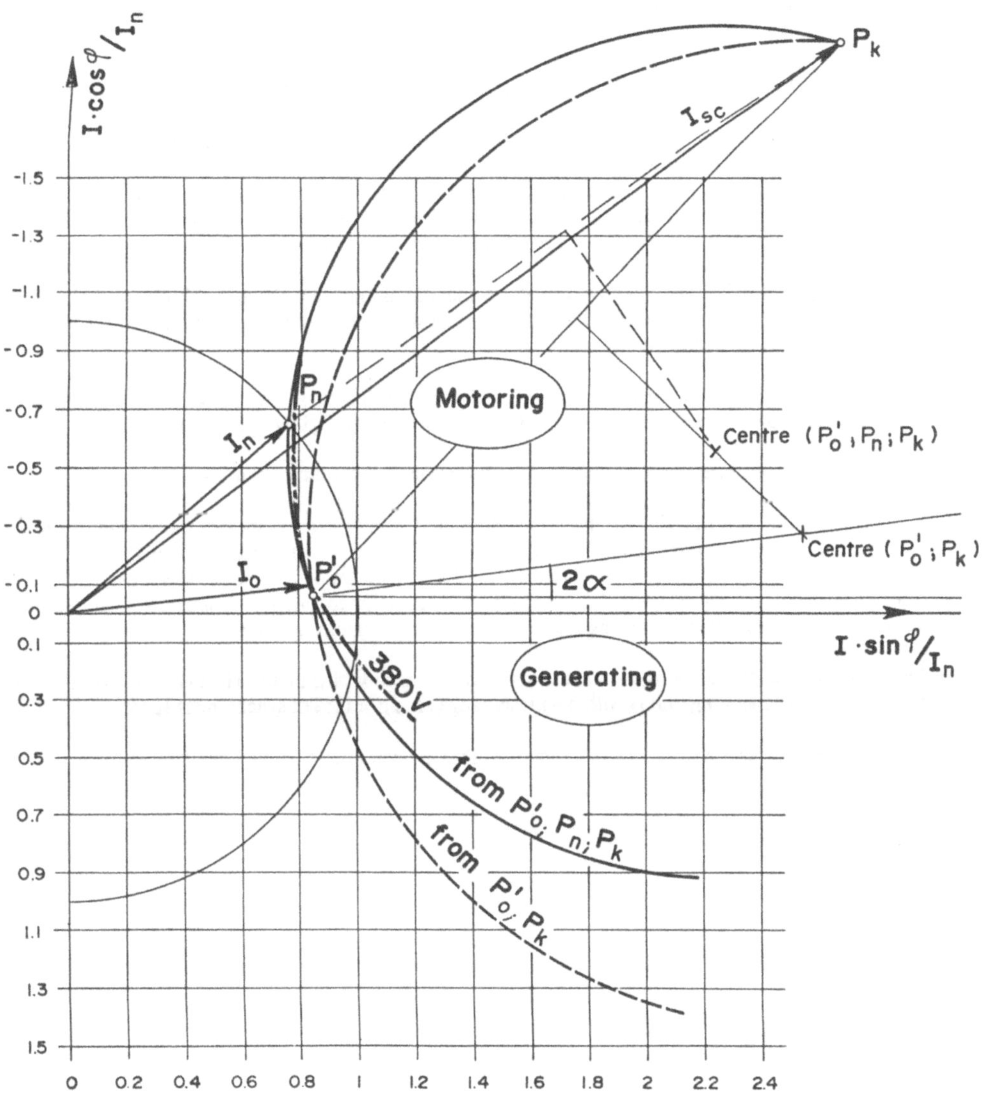

FIGURE 4.9 :
Performance of an induction machine (P_n = 2.2 kW) predicted by no-load and blocked rotor tests in comparison with actual performance

For more details on how to plot the circle diagram from no-load and blocked-rotor test data see Appendix C.

5. OPERATION AND CONTROL

5.1 Operation in Parallel with a Large Grid

5.1.1 Start-up

Unlike the synchronous generator, the IMAG need not be synchronized to the voltage, frequency and phase angle of the grid. The following simple procedure applies for small induction machines (<500 kW):

- the turbine control valve or the guide vanes are opened slowly; the generating unit speeds up; when the unit reaches about synchronous speed, the circuit breaker can be closed, connecting the IMAG to the grid;

- at undersynchronous speed the IMAG operates as a motor and, similar to motor start-up, the current drawn from the grid will exceed nominal current. That is why turbine speed should be brought **as close as possible to synchronous speed** before connecting the unit to the grid; however, even at synchronous speed, a relatively high transient current peak (the so-called in-rush current) will occur when switching which is due to the fact that the IMAG represents practically a short circuit for the grid (the stator is not yet magnetized and shows little reactance to the current). Due to its short duration (about 10 ms), this in-rush current has no negative effects for IMAGs parallelling to a large grid;

- increasing the turbine gate opening will drive the generating unit above synchronous speed, i.e. into generating mode; the unit will now start to deliver power to the grid. The turbine gate can be opened according to the available water;

- the capacitors, which may be connected in parallel to the generator (to improve the power factor), are either connected automatically together with the circuit breaker or slightly afterwards.

The procedure is reversed when the unit needs to be disconnected from the grid:

- turbine flow is reduced until the generating unit operates as a motor; the circuit breaker and simultaneously the capacitors may then be disconnected. (In case of emergency, the main switch must be disconnected immediately, despite the fact that the turbine-generator unit might speed up.) Note that the capacitors must not remain charged but should be short-circuited through resistors to avoid electricians working on the installation from being electrocuted by a discharging capacitor.

More sophisticated installations use an automatic grid connection; the main switch is automatically closed if the speed of the generating unit is within a preset range of e.g. ±10% of synchronous speed.

Figure 5.1 below shows a schematic diagram of an IMAG operating in parallel to a large grid.

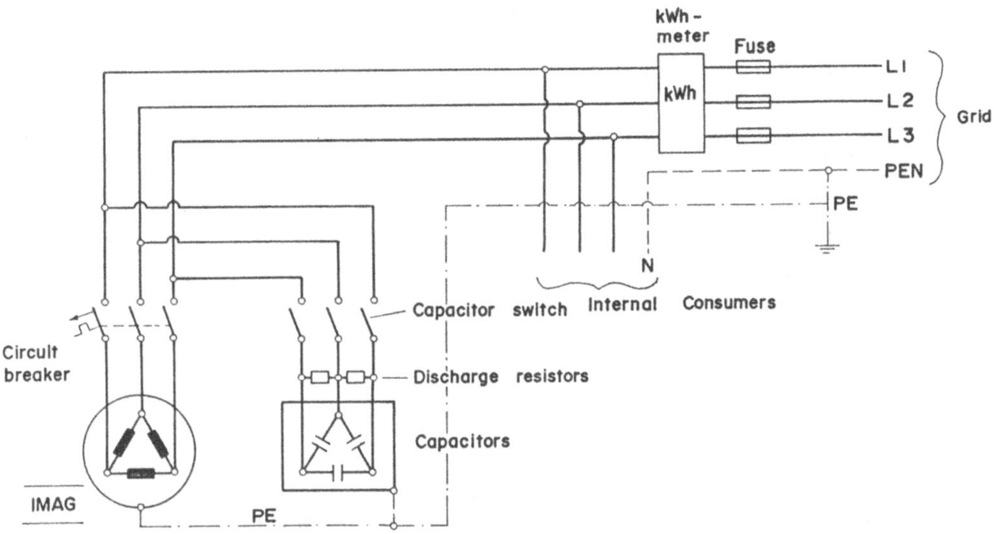

FIGURE 5.1 :
Schematic diagram of an IMAG operating in parallel to a large grid

5.1.2 Compensation of the Reactive Power Demand

As mentioned earlier (Section 3.3.1), the induction motor cannot by itself produce the excitation current to establish its magnetic field. In parallel operation, this reactive power might be supplied by the grid. Nevertheless, capacitors are usually connected in parallel to the IMAG to improve its power factor; i.e. the capacitors and the stator winding of the IMAG form a resonance circuit where the reactive power oscillates only between capacitors and induction machine. In this way, the grid is not loaded with reactive current and transmission losses can be reduced. Electricity companies usually demand a minimum value of power factor of the generating unit (typically 0.9); if a higher reactive power is drawn from the grid, the electricity company will measure this demand and charge for it.

The capacitance required by the IMAG may be determined by the formula developed in Appendix A:

$$\boxed{C_{\Delta \text{ per phase}} = \frac{\Delta Q}{3\,U^2\;2\,\pi\,f}} \qquad (5.1)$$

where U = line voltage (if capacitors are connected in delta as shown in Figure 5.1 above)
f = grid frequency and
ΔQ = reactive power to be compensated

If the capacitors are connected in star (Y), the capacitance required per phase is three times larger than according to Formula 5.1.

$\Delta Q = Q_{\text{reactive power before compensation}} - Q_{\text{reactive power after compensation}}$

$$\boxed{\Delta Q = P_{el_g\,eff}(\tan\phi_1 - \tan\phi_2)} \qquad (5.2)$$

where ϕ_1 = phase angle between phase voltage and current before compensation and ϕ_2 afterwards; at full compensation $\cos\phi_2 = 1$ and $\tan\phi_2 = 0$.

The problem is that the power factor $\cos\phi_1$ when generating usually differs from the power factor when motoring (see also section 3.3.2 above). While the motor power factor of an IMAG is usually known from the rating plate or manufacturers brochures, $\cos\phi$ for generating must be estimated (or measured). Figure 4.4 above indicates the relationship between motor and generator $\sin\phi$. This diagram has been plotted from test results using unsaturated standard motors at full load and nominal machine current.
The drop of power factor (i.e. increase of reactive power) when changing from motoring to generating largely depends on the design of the IMAG (rotor bar construction, saturation limit).

The relationship given in Figure 4.4 will enable accurate sizing of capacitors for non-saturated motors only; for saturated ones an accurate calculation of the required capacitors is not possible. It is therefore advisable to undertake the final selection of the capacitors on site after installation of the IMAG. The size of the capacitors must be adjusted until the required power factor is attained. Since instruments measuring the power factor are not standard equipment for small hydropower, the following instruments might be used to calculate the power factor (see also Appendix D: Measuring techniques and required equipment):

- Wattmeter (or kWh counter) measuring active power (or energy) delivered to the grid (the Wattmeter may be a portable instrument while a kWh meter is usually standard equipment for plants operating in parallel (measuring electricity delivered for reimbursement)

- with ammeter and voltmeter for each phase (standard monitoring equipment for any hydropower plant), total or apparent power can be determined:

$$S = \sqrt{3}\, U_{line}\, I_{line} \qquad (5.3)$$

Using the relationship $\cos\phi = P / S$, the power factor can be calculated.

5.1.3 Example on power factor correction for parallel operation

The IMAG selected in Section 4.3.5 for the small hydropower plant of 11 kW turbine output should be connected to the national grid. The electricity utility demands a minimum value of power factor of 0.9. Select the required capacitors.

Capacitors for power factor correction are required for the effective power only; the power factor at the effective operating point can be drawn from the diagram in Section 4.3.5 above:

$\cos\phi_g\ \text{eff} = 0.695$

We can now calculate the reactive power demand of the IMAG at the effective electrical power of the plant (P_{el_g} effective = 9.5 kW).

$$Q_{total} = P_{el_g\ effective}\tan(\arccos\phi) = 9.5\tan(\arccos 0.695) = 9.85\ \text{kVar}$$

reactive power to be supplied by capacitors for a power factor required by the electricity company $\cos\phi_{req} = 0.9$:

$$Q_c = \Delta Q = Q_{total} - P_{el_g\ effective}\tan(\arccos\phi_{req}) = 9.85 - 9.5\tan(\arccos 0.9) = 5.2\ \text{kVar}$$

Capacitors in delta connection per phase:

$$C_{\Delta\ per\ phase} = \frac{Q_c}{3\,U^2\ 2\,\pi\,f} = \frac{5200}{3\ 380^2\ 2\ \pi\ 50} = 38 * 10^{-6}\,F = 38\,\mu F$$

5.1.4 Compensation at part-load operation

An additional problem arises at times when the machine cannot be operated at nominal power due to a lack of water. The power factor determined from Figure 4.4 and the corresponding capacitance selected will be correct for nominal output; for part-load operation, the capacitance selected might be too large despite the fact that the power factor diminishes too with decreasing power. The (overcompensated) IMAG would therefore supply capacitive power to the grid at part load producing additional losses on the supply line (higher line current). This relationship is presented in Figure 5.2 below for a 3 kW motor. The maximum power output when generating (P_{el_g} max = 2.4 kW) and the power factor curve at part load have been obtained from test results. The example shows that for this particular case, a fixed capacitance of 19 μF per phase (delta connected) is just enough to correct the power factor to the required 0.9 at full load but will not overcompensate the IMAG down to 25 % part load.
For larger installations, especially with long supply lines, an automatic adjustment of the power factor using varying capacitors might be considered. Smaller plants are usually equipped with capacitors of fixed size.

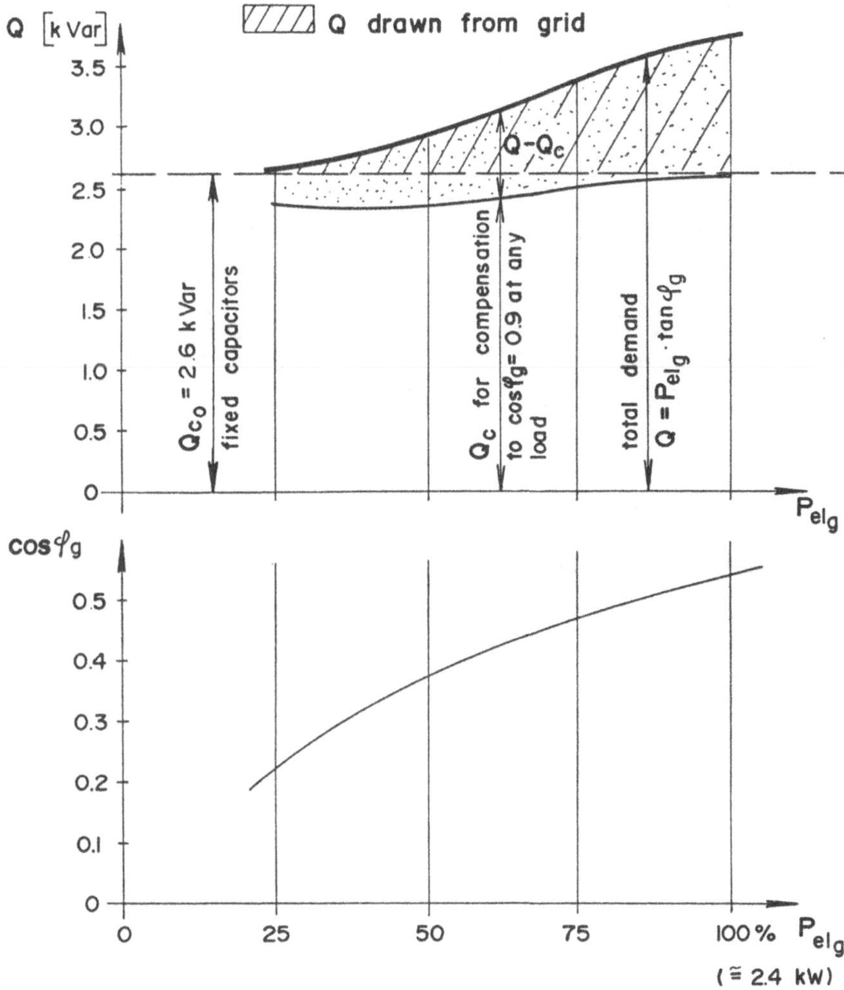

FIGURE 5.2 :
Power factor compensation of an IMAG in parallel operation with a grid (for a 3-kW IMAG)

5.1.5 Safety and Control Equipment for parallel operation

The electrical equipment required for a micro-hydropower plant can be divided into two main parts, namely, the switchgear, containing the equipment for system connection and power distribution, and the monitoring and control equipment, which do not carry high currents and voltages but are nevertheless essential for the proper operation of the plant. A schematic overview of the electrical components of an MHP using an IMAG is presented in Figure 5.20 at the end of this chapter. The protection devices from the point of view of safety are treated in more details in Section 6.5.

Basically, there are two functions to be fulfilled by these devices:

- to monitor and protect the grid
- to monitor and protect the MHP itself (IMAG and turbine)

Grid Protection

In case of irregularities or failure of the grid, the generator working in parallel to it must be disconnected: this is in the first place required by the electricity company in order to avoid that the IMAG continues to supply electricity into a system which does no longer meet the quality standards (protection of the consumers). To monitor the state of operation of the grid, the electricity boards usually demand the following meters and the corresponding protective devices:

a) **over- and undervoltage protection** (one per phase), usually composed of voltage relays which trip the main circuit breaker if the grid voltage of any one phase exceeds or falls below the preset limits during a certain time;

b) **over- and underfrequency protection** (required only for MHPs of a certain size, regulations differ slightly from country to country). Instead of monitoring grid frequency, electricity boards sometimes allow to monitor the rotational speed of the generating unit which is indirectly governed by the grid frequency. This has the advantage that the same device may also be used to activate the overspeed protection of the turbine (see below). Omitting costly frequency meters and trips is especially important for small plants where cost is a major consideration.

These devices are usually of the automatic reset type; i.e., the IMAG is re-connected automatically to the grid as soon as the fault has been cleared.

Internal Protection and Control

Electric faults and irregularities may be caused by the IMAG itself or its ancillary equipment (capacitors, etc.). In such a case, the protection devices must shut down the plant and alert the operator. These devices allow only a manual reset. Some of the devices are also a back-up safety precaution for the grid protection devices (i.e. the machine is disconnected even if the main circuit breaker fails).

a) **overload and short-circuit protection**
Fuses or circuit breakers on all three phases must be installed for short-circuit protection. Overload protection may be composed of either a bimetal (thermal) or an electromagnetic device sensing the overcurrent.

b) **overvoltage protection** devices are used to shut down the generator when changes in voltage exceed a preset limit (possibly caused by overspeed of the IMAG, see also paragraph c) below). Overvoltage relays are normally used for this purpose.

On opening the circuit breaker of an IMAG in parallel operation, the generated voltage may not immediately collapse if the supply of magnetizing power (for its excitation) is maintained by the capacitors; this phenomenon is referred to as self-excitation of the IMAG in parallel operation. In fact, voltage will rise since the capacitors of fixed size deliver a constant magnetizing current for operation under load. At no-load, the same magnetizing current must be absorbed by the IMAG which is only possible if voltage rises (see Figure 5.4 below). Thus, it is essential that capacitors are disconnected immediately upon an opening of the circuit breaker.

Even with disconnected capacitors, the voltage at the terminals of the IMAG will persist for some time though slowly decaying according to an exponential law. The following formula may be used to estimate the voltage decay of an IMAG:

$$\boxed{U_{(t)} = U_n \; e^{-t/\tau}} \qquad (5.4)$$

where U_n = nominal voltage

t = time elapsed in seconds after opening the circuit breaker

τ = time constant in seconds $\tau = \dfrac{x_m + x_{\sigma r}'}{2 \pi f \; r_r'}$

x_m = magnetizing reactance; order of magnitude for P >10kW: 2.5 - 4.0 per unit

$x_{\sigma r}'$ = leakage reactance rotor 0.08 - 0.12 p.u.

r_r' = rotor resistance 0.01 - 0.08 p.u.

f = frequency in cycles per second

Definition of 'per unit' values: $x = X / Z_n$ where $Z_n = U_{phase} / I_{phase}$

For a frequency /speed of the generating unit of 50 Hz (or n_s = 1500 rpm for a four pole machine) we obtain a time constant of $\tau = (4.0 + 0.12)/(2\pi\ 50 * 0.01) = 1.31$ s

One second after breaker opening, the voltage would have dropped to $U_{(t=1)} = U_n\ e^{-1/1.31} = 0.47\ U_n$.

Therefore, circuit breaker reclosure must be delayed until the IMAG terminal voltage has practically disappeared; too rapid breaker reclosure may result in severe machine shock because the remaining generator voltage may be out of phase with grid voltage. The effect is similar to synchronizing a synchronous generator out of phase. The formula above may be used to determine the minimum period of time before a manual or automatic reclosure of the circuit breaker can be effected.

c) <u>overspeed protection</u> is generally required in conjunction with a general grid failure or circuit breaker opening. The sudden loss of load, such as with a breaker opening, causes rapid generator-turbine acceleration (IMAGs have in general lower rotor inertia than synchronous generators and therefore speed up faster; this tendency is increased if a pump is used as turbine; PAT). Generally, it is more economic to design the turbine-generator unit in such a way that all components can operate at runaway speed than to prevent turbine overspeed.

Turbine shut-down devices which respond fast enough to avoid overspeed of the generating unit are usually complicated and may cause waterhammer problems in the hydraulic system. An exception is the jet-deflector of impulse turbines (Pelton, Turgo) which diverts the flow of water away from the turbine. The jet deflector can be actuated by a spring or a counterweight lever. During normal operation, the deflector is kept out of the jet by a magnetic switch or a solenoid. Upon grid failure or other irregularities, the circuit breaker opens and disrupts the magnetic switch of its electricity supply; the spring or counterweight lever is released pulling the deflector across the jet and the turbine slows down. The solenoid could be controlled also by other relays sensing irregularities of the grid or generator operation (e.g., overvoltage relay, tacho, etc.).

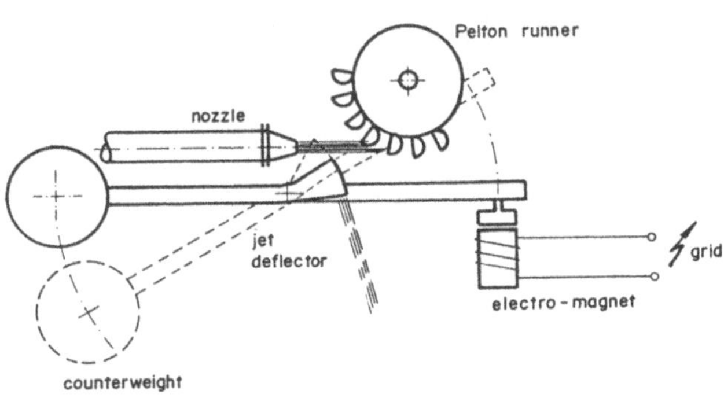

FIGURE 5.3 :
Overspeed protection by means of a solenoid, counterweight lever and jet deflector of a Pelton turbine

In addition to mechanical hazards (centrifugal forces on rotating parts), the effects of overspeed may have far-reaching consequences on the voltage of the IMAG. As explained under b) above, generator voltage still persists if, after disconnecting the load/grid, the capacitors remain connected to the IMAG (self-excitation). In such a case, voltage will increase from the normal operating point a) to no-load voltage, point b) Figure 5.4 below. While the demand of magnetizing current of the IMAG decreases with the overspeed of the unit, the supply of reactive power by the capacitor battery goes up due to the increased frequency (see also section 3.3.5 , Figure 3.16). Voltage must therefore increase (point c in Figure 5.4 below) until a new equilibrium between supplied and absorbed reactive current is achieved. For the example presented in Figure 5.4 we have assumed a runaway speed of 1.2 times rated speed corresponding to 60 Hz instead of 50 Hz. Figure 5.4 shows that the final voltage (point c) is at 565 V which is 1.4 times the nominal value of 400 V. Hence, the overvoltage ratio is larger than the overspeed ratio.
This example shows that it is in any case necessary to **disconnect the capacitors** as soon as the IMAG is no longer under load. Failing to do so will invariably lead to damage on stator windings, capacitors and monitoring instruments and may endanger people.

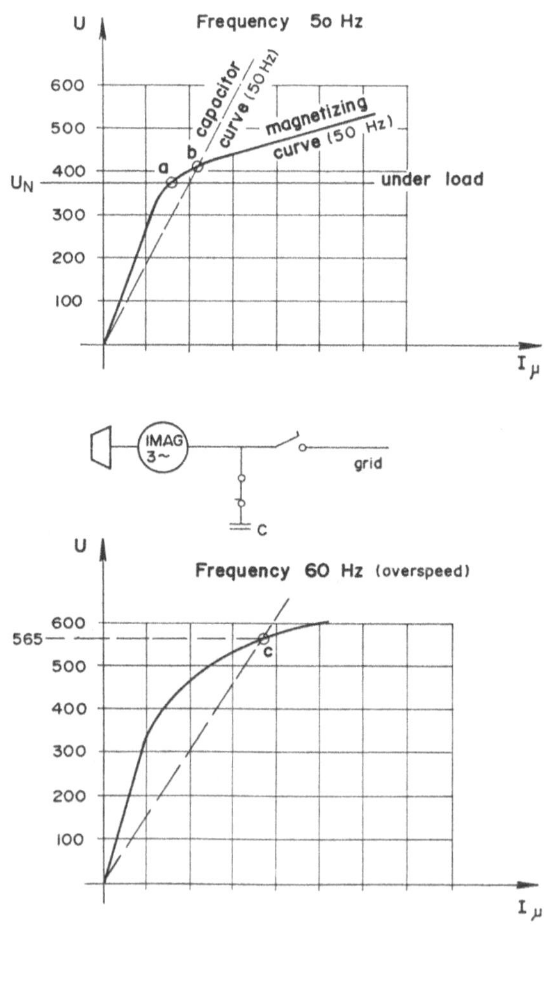

FIGURE 5.4 :
Increase of voltage due to overspeed of the IMAG at open circuit (runaway speed)

Since a turbine shut-down is often required upon grid failure, the hydraulic control devices (valves or guide vanes) cannot be actuated using grid power. Various forms of turbine shut-down devices and their actuators exist for small hydropower plants:

- the simplest form is the manual control/closure of turbine gates or guide vanes; since MHPs working in parallel to the grid are not continuously supervised by an operator, manual control is only possible if the generating unit can stand runaway speed for a long period (several hours) without any damaging effects (IMAG of course de-excited);
- servo-motors/linear actuators (DC) moving turbine guide vanes or electrical motors (DC) closing penstock control valves can be supplied by a rechargeable battery (12 or 24 V);
- many turbines are traditionally equipped with oil-hydraulic governors controlling the guide vane opening in function of the water level in the forebay. Though not recommended in conjunction with IMAGs (costs!), oil-hydraulic governors have the advantage of being driven by flat belts coupled to the turbine; the oil pressure is therefore maintained continuously and turbine closure is possible also after grid failure. (Another drawback of oil-hydraulic governors apart from their costs and complexity is their inherent risk of polluting the water which is especially serious in MHPs incorporated in water supply systems.)
- an alternative to oil pressure is the use of a pneumatically actuated penstock valve or guide vane (compressor serving an air-pressure reservoir);
- the pressure energy from the penstock in conjunction with a counterweight or a spring is another possibility to shut down the turbine in case of emergency or grid failure.

d) **reverse-power protection** is recommended for any IMAG working in parallel to a grid. If the speed of the generating unit becomes undersynchronous due to a lack of water or pressure on the turbine, the IMAG will start to operate as a motor and draw power from the grid. A reverse-power relay will sense a reverse flow of power and disconnect the IMAG.

e) **phase-imbalance protection** is recommended in three-phase systems; if the total current is no longer evenly distributed over the three windings because of a phase failure or unbalanced loads on the grid, the IMAG must be disconnected to prevent overheating and subsequent damage. The overcurrent protection (see above) per phase is not sufficient to sense a phase imbalance or failure of one phase; this is especially true for fully-compensated IMAGs. Phase imbalance relays are used for this purpose which may trip at 10 to 20% phase imbalance on the grid.

f) **supervision of winding and bearing temperature** may be considered in larger installations where damage on the unit could lead to considerable investments; the need for such protection must be evaluated on its cost-effectiveness from case to case. Some motors have thermal sensors incorporated on the windings as standard equipment including their connection to the terminal box.

Optimization of energy production - Water level control

For run-of-the-river MHPs with large variations in stream flow between wet and dry season, water level control is advisable. The design flow of the plant is larger than the minimum flow. When the available flow drops below design discharge of the turbine during the dry season, the turbine opening must be reduced in order to maintain normal water level in the forebay. This can either be done manually or automatically. The automatic control uses a mechanical float or pressure gauges (inductive or piezo-electric gauges) placed in the forebay to sense the variations of the water level; the turbine opening is adjusted to maintain this level between a maximum and a minimum value
(see Figure 5.5 below).
Water level control for MHPs using an IMAG may not prove to be as cost-effective as for plants with e.g. purpose-built induction generators since efficiencies of IMAGs drop fairly rapidly at part load; the increased energy production of a purpose-built induction generator accumulated over the whole service life of the machine may therefore offset its price disadvantage as compared to the IMAG. This is especially true for plants operating for extended periods at part load. A careful evaluation of the different generator options (usually IMAG or purpose-built induction generator) is an essential part of the MHP design when parallel operation to a large grid is possible (for the determination of the optimum design flow and the cost-benefit analysis see Chapter 7. Economic considerations).

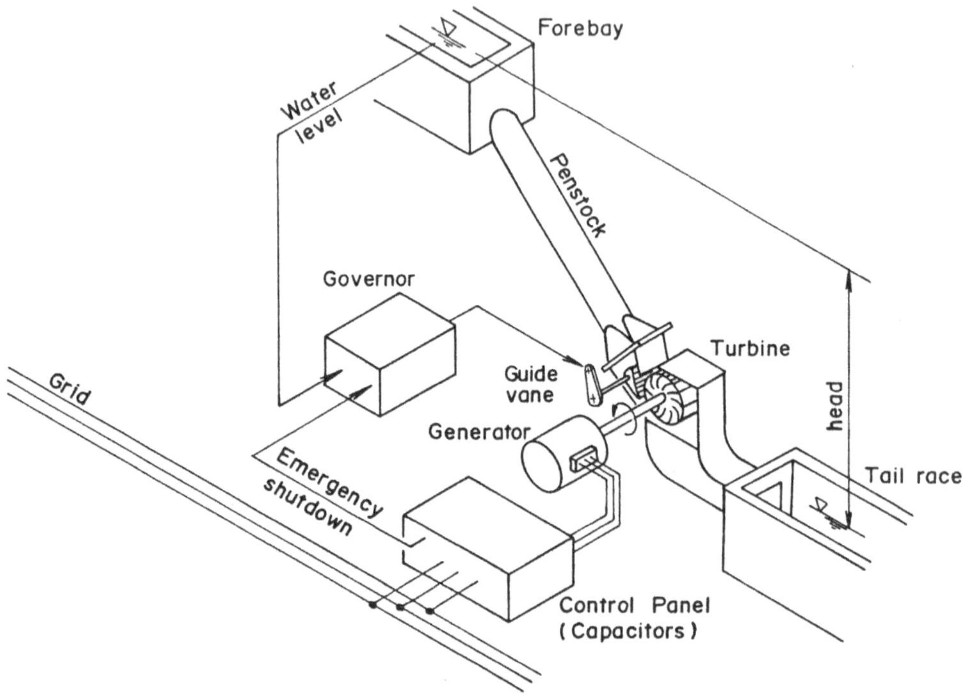

FIGURE 5.5 :
Water level control of an MHP in parallel operation

An overview of the necessary monitoring, control and protection devices is provided in Table 5.21 at the end of this chapter.

5.2 Operation in Parallel with a Synchronous Generator (Minigrid)

The following rules apply if an IMAG is to be operated in parallel with a synchronous generator in an isolated grid (minigrid):

- stable operating conditions can be achieved if the rated apparent power of the synchronous generator is at least twice the rated apparent power S of the IMAG;

- capacitors need not necessarily be connected to the IMAG since reactive power could be supplied by the synchronous generator operating in overexcited mode;

- the monitoring and protection equipment required is similar to operation in parallel to a large grid.

5.3 Stand-alone Electricity Generation

5.3.1 General

Stand-alone operation of an MHP is characterized by the condition that generated power must be equal to consumed power at any instant. In an unregulated plant supplying varying loads, this equilibrium is established by the machine itself through varying frequency and voltage. If the power demand drops below rated output of the generator, speed and voltage increase until all generated power is absorbed by the consumers; when all consumers are disconnected, the generating unit will reach runaway speed, where all the hydraulic input power is dissipated in hydraulic and mechanical losses (efficiency = 0). When overloaded, the machine reacts with a drop of speed and voltage which reduces both output power and consumed power (due to reduced voltage). This will lead (though not always) to a new equilibrium on a lower power level.
These changes in voltage and frequency are undesirable from the point of view of the consumers. The following table indicates the effects of changing voltage and frequency on different loads.

TABLE 5.6 :
Effects of frequency and voltage variations on various electrical loads

Electrical load	*Decreasing Frequency*	*Increasing Frequency*
heating loads (without ventilator)	- not affected	- not affected
lighting loads - incandescent bulbs	- not affected	- not affected
- fluorescent lamps	- coils heat up and may burn due to reduced reactance and higher currents at low frequency ($f_{min} = f_n - 5$ %)	- increases reactance of coils which allows only reduced current to flow
transformers	- windings may burn due to reduced reactance and higher currents at low frequency	- increases reactance of coils which allows only reduced current to flow
motor loads	- as transformers - reduces speed and affects output torque and power development of driven equipment	- as transformers - increases speed and affects output torque and power development of driven equipment
refrigerators - absorber type	- not affected (similar to heaters)	- not affected (similar to heaters)
- compressor type	- as motors	- as motors
electronic devices (radio, TV, tape, computer)	- as transformers	- as transformers

Generally, slight increase of frequency is acceptable for most consumers, whereas decreasing frequency should be avoided.

Electrical loads	*Decreasing Voltage*	*Increasing Voltage*
heating loads (without ventilators)	- generated heat /power varies with the square of the voltage variation; requires more time to do the same work (cooking, etc.)	- heating element may burn due to excess heat generated
lighting loads - incandescent bulbs - fluorescent lamps	 - decreases light output but improves service life - lamp may not start or may flicker during operation	 - decreases service life considerably even at small increase of voltage
transformers	- not affected (power decreases)	- transformer heats up because of a higher magnetizing current (it operates at a higher saturation level; see Figure 3.14)
motor loads	- torque of induction motors varies with the square of the voltage; motor may not be able to start the driven equipment with the reduced torque and may burn due to overcurrents	- as transformers
refrigerators - absorber type - compressor type	 - as heating loads - as motors	 - as heating loads - as motors
electronic devices (radio, TV, tape, computer)	- accept fairly high voltage variations without damage	

An IMAG always requires an external source of reactive power to establish its magnetic field. In isolated operation, this source is composed of a battery of capacitors connected in parallel to the IMAG. Generally, the reactive power demand varies with the load put on the IMAG. This implies that the size of the capacitors need to be adjusted with varying load in order to maintain constant voltage. The control of an isolated system using an induction generator becomes therefore more difficult than with a synchronous generator which usually incorporates an automatic voltage regulator (AVR). The following parameters determine the frequency and the voltage of an IMAG in stand-alone electricity generation; controlling these will thus ensure a good quality of the generated electricity:

- speed of the generating unit (IMAG and turbine)
- power demand of the consumers
- value of the installed capacitance parallel to the IMAG

A major advantage of the IMAG is its lower cost as compared to synchronous or purpose-built induction generators. This cost advantage could be lost to a large extent if the operating and governing concept of the IMAG did not adopt the same low-cost philosophy. Full-fledged speed (e.g., using a flow-control governor) and voltage control in conjunction with an IMAG does not make sense since these devices would cost ten times as much as the IMAG itself (for a 10 kW machine). By allowing some losses on the quality of the generated electricity, simplified approaches to governing of IMAGs are possible. The following options are generally applied for IMAGs in stand-alone electricity generation:

- ungoverned plant in conjunction with constant load
- hand-regulated speed governing (turbine flow) and voltage control using stepped capacitors
- load control (induction generator controller IGC sensing voltage and leaving frequency to fluctuate)

(these options have been compared with other concepts in view of their voltage and frequency variations in section 2.3.5 above).

5.3.2 The Ungoverned Plant at Constant Load

The circuit contains no switches except the main switch at the power house. Closing this switch supplies electricity to all consumers simultaneously; similarly, turning off appliances is only possible if the IMAG is stopped, closing down the whole system. This concept is usually applied for village lighting and heating devices. Note that the individual loads must not vary their power demand during operation (e.g., motors driving machine-tools etc.) otherwise constant load on the generating set could not be ensured without a governor.

Start-up

Starting is relatively difficult because the reactive power demand is different before and after connecting the load. The battery of capacitors should therefore come in two separated sets: the first set called excitation capacitor should provide just enough reactive power for the excitation of the IMAG but should not produce overvoltage when turbine flow is increased. Start-up would therefore be as follows (see Figure 5.7 below):
- the turbine gate is opened so that the generating unit reaches about nominal speed;
- with the aid of the first set of capacitors, self-excitation of the IMAG is brought about (see also section 3.3.4 above)
- just before closing the circuit breaker, turbine flow is increased in preparation of full power delivery;
- the unit speeds up, increasing the terminal voltage of the IMAG; the first set of capacitors is chosen in such away that this overvoltage does not exceed the permissible voltage of the winding (typically 1.3 U_n);
- on closing the main switch, the load and the second set of capacitors are connected; these capacitors supply the additional reactive power required to establish nominal voltage under load.

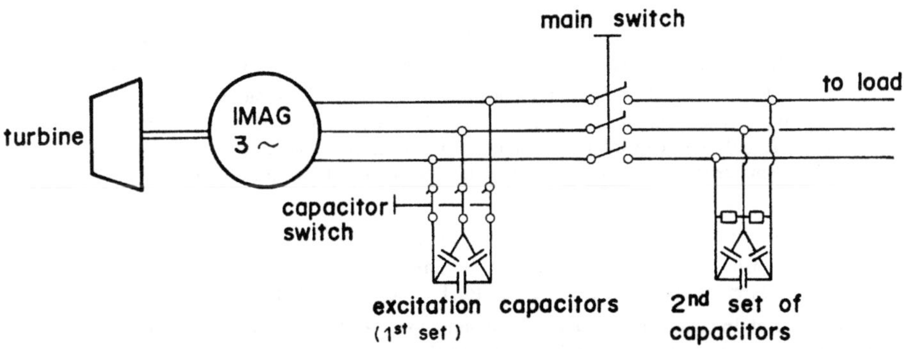

FIGURE 5.7 :
Start-up of an IMAG in isolated operation using two separate sets of capacitors

An elegant way to start up an IMAG is the use of over- and undervoltage relays, which are in any case a recommended protection device for MHPs using IMAGs. The electrical layout of this system is presented schematically in Figure 5.8. The IMAG is started as explained above but this time the main switch is closed right at the beginning. However, the undervoltage relay prevents the flow of power to the consumers before excitation of the IMAG is completed; once voltage exceeds, say 120 % of nominal voltage, the relay closes, thus, connecting the consumers. This procedure is easier from the operator's point of view since he can open the turbine gate without having to supervise the electrical circuit. In case of electrical faults on the consumer side, voltage cannot go beyond the preset trip voltage value of the overvoltage relay (typically 1.3 times nominal voltage). Note that tripping this relay causes the generating unit to accelerate to runaway speed.

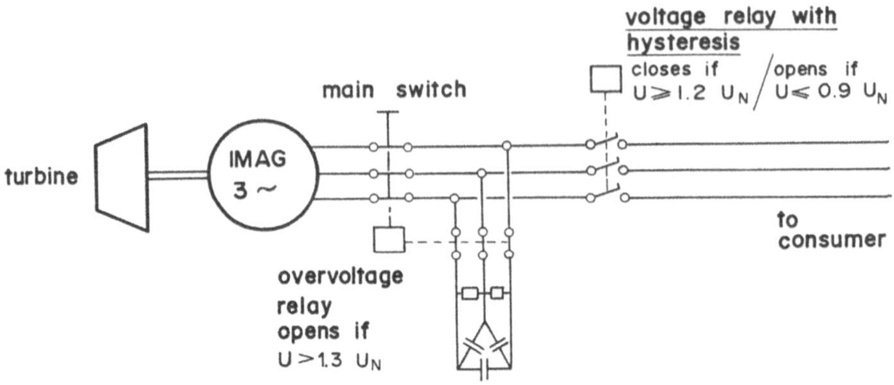

FIGURE 5.8 :
Start-up of an IMAG in stand-alone operation using under- / overvoltage relays

Sizing Capacitors

Selecting capacitors for the IMAG in isolated operation basically follows the same procedure as presented in section 5.1.2 Parallel Operation, with the exception that the capacitors must be sized to supply total reactive power or $\cos\phi = 1$. For isolated operation the selection of capacitors is not merely a question of power factor correction but in fact determines the operating point of the IMAG. For a constant electrical load and a given machine, there is only one capacitor size (and turbine opening) which will yield nominal voltage and frequency. This statement is examined in more details using the following example.

__Example:__ A motor of 2.2 kW rated power ($P_{el_m} = P_n/\eta_m = 2.2/0.83 = 2.65$ kW) is to be used as generator in stand-alone electricity generation (unregulated, constant load). Using the diagrams of Chapter 4, we have determined its maximum power output as generator P_{el_g} to be 0.82 times P_n or 1.8 kW. The IMAG is to supply constant lighting and heating loads corresponding to 1.8 kW i.e full load operation.
Applying the same procedure as under 5.1.2, we can determine the reactive power demand of the IMAG. With a motor power factor of $\cos\phi_m = 0.82$ at nominal load we can calculate the reactive power demand when motoring $Q_m = P_{el_m} \tan\phi = 2.65 \tan(\arccos 0.82) = 1.85$ kVar
Figure 4.4 provides (for a rated motor power of $P_n = 2.2$ kW) a ratio of reactive power demand between generating and motoring ($\sin\phi$ - ratio) of:

$$\sin\phi_g / \sin\phi_m = 1.45$$

Hence, reactive power demand when generating: $Q_g = 1.45 * 1.85$ kVar $= 2.68$ kVar and the corresponding capacitors for a three-phase / 380 V distribution (capacitors in Δ):

$$C\Delta_{\text{per phase}} = 2680 * 10^6 / (3 * 2\pi\, 50 * 380^2) = 20\ \mu F$$

These capacitors are divided into two sets, the excitation capacitors and the load capacitors; The value of the excitation capacitor is usually taken as the motor nominal load capacitance (which corresponds approximately to no-load capacitance when generating):

$$Q_{excit} = Q_m = P_n/\eta_m * \tan(\arccos\phi_m) = 2.2/0.83 * \tan(\arccos 0.82) = 1.85 \text{ kVar}$$

$$C_{excit} \text{ per phase} = 1850 * 10^6 / (3 * 2\pi\, 50 * 380^2) = 14\ \mu F$$

As mentioned earlier, the ratios of Pel_g/P_n and $\sin\phi_g/\sin\phi_m$ are subject to quite a large dispersion depending on the type of motor selected (saturation limit, rotor bar construction). The operating conditions calculated above may be different from the actual performance of the machine on site. After installation of the IMAG there are only two parameters which can be adjusted to establish the voltage and frequency required by the consumers:

- turbine speed and output by means of the turbine gate opening (provided the available stream flow allows to do so), and
- reactive power supply through capacitors.

However, the adjustment of speed and the verification of the correct voltage and frequency requires at least the following instruments:

- voltmeter and - ammeter } these two should be permanently installed in MHPs of whatever size
- frequency meter (could be a portable instrument, to be used for adjustment of capacitors during commissioning of the plant only)

The following procedure applies:

- connect capacitors as calculated above and start up as described at the beginning of this section

- monitor voltage and adjust turbine flow/opening to arrive at nominal voltage

- the IMAG will adjust itself to that speed/frequency which will yield the reactive power needed for the imposed operating point; this frequency will most probably not correspond to the nominal frequency (50 or 60 Hz) of the consumers but will be slightly above or below it according to the capacitance connected.

Figure 5.9 below shows this behaviour using the circle diagrams for three different capacitor sizes and the corresponding frequencies; note that the electrical output power (= load) remains constant despite the varying frequency (this is true for pure resistive loads only; for inductive loads see below). However, the mechanical input power from the turbine must be adjusted according to the varying reactive power for different frequencies/capacitor sizes (internal losses increase with increasing reactive power; hence, efficiency of the IMAG drops)

- to find the new capacitor size required to establish nominal frequency f_n, the following formula applies (see analogy with resonance frequency, Appendix A, section 7.10):

$$\boxed{C_n = C\left(\frac{f}{f_n}\right)^2} \tag{5.5}$$

From Figure 5.9 we can see that the frequency with 60 μF would be 52 Hz. To bring it to a nominal frequency f_n of 50 Hz, the capacitor size should be adjusted to the following value:

$C_n = 60\,(52/50)^2 = 65\ \mu F$

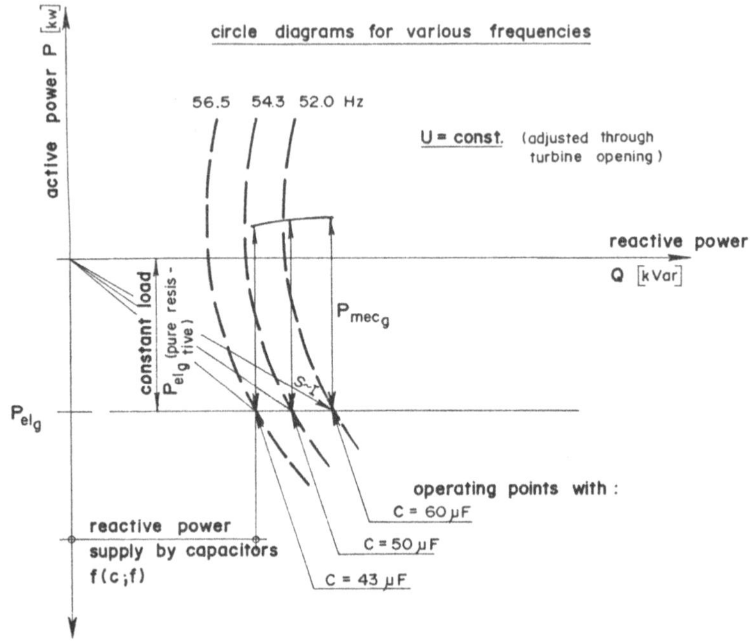

FIGURE 5.9 :
Adjusting frequency of an IMAG in isolated operation by varying capacitor size and turbine output

Adjusting the capacitance of the IMAG in case of inductive loads is not as straightforward as explained above for pure resistances. If the electrical load contains inductive appliances such as fluorescent lamps, the power factor of the load should first be brought to unity (capacitor size determined with nominal frequency) before starting measurements with the IMAG. Furthermore, formula (5.5) will give only approximate values since the reactive power of the load will be changing with varying frequency of the generated electricity; the final capacitance C_n for the IMAG will therefore have to be found by trial and error.

Generally, a frequency slightly above the nominal value has no detrimental effects for most appliances used in an unregulated system (heaters, fluorescent and incandescent lamps, electronics). Frequency variations between ±0 and +5% are usually permissible.

De-excitation

A particularity of the IMAG in isolated operation must be mentioned here. Overloading a synchronous generator causes an increased current to flow. The AVR tries to maintain the nominal voltage; without an overload protection, the generator windings will heat up and finally burn due to the overcurrent. On the other hand, the unregulated IMAG with fixed capacitors (excitation) in isolated operation is inherently overload safe. The maximum electric power P_{el_g} to be generated by the IMAG is determined by the excitation capacitor connected. Increasing load will cause both voltage and frequency to drop. The magnetizing power demand of the IMAG increases whereas the reactive power output of the capacitors decreases with the dropping frequency (see also Section 3.3.5 above). Without adjusting capacitor size, the voltage of the IMAG will decrease further and excitation will finally collapse. Figure 5.10 below shows the voltage and current variations for an increasing load up to system breakdown for different sizes of capacitors (note that frequency has been kept constant in order to clarify the effect). After de-excitation, the IMAG will not recover; the consumers must be disconnected and the IMAG be magnetized as explained under the heading Start-up. System breakdowns under load may wipe out the remanent voltage of the iron core which allows excitation of the IMAG. Re-magnetizing the core can be achieved by connecting a battery (6 or 12 V, allowing about nominal current to flow) for a few seconds to the stator winding at standstill of the machine.

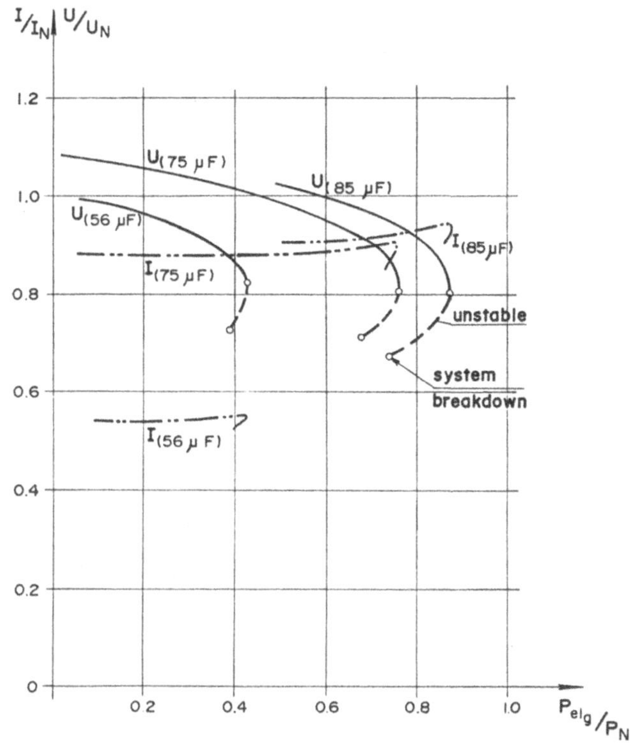

FIGURE 5.10 :
Development of voltage and current for increasing load and limit of de-excitation (at constant frequency) for a 3 kW motor (from test results)

Protection Required

An overview of recommended monitoring and protection equipment for unregulated constant load installations is given at the end of this chapter.

5.3.3 The hand-regulated plant with stepped capacitors

When electrical loads in a stand-alone system are changing, some form of speed control is usually necessary to match at any instant the power supplied by the generator with the power demand of the consumers. This **control of active power** is conventionally done by mechanical speed governors acting on the turbine opening. For low-cost applications using an IMAG, a mechanical governor is often too expensive and hand-regulation may be a more appropriate solution.
The previous section has shown that some form of **reactive power control** is also necessary in order to maintain voltage within acceptable limits, even more so if inductive loads such as induction motors are connected to the system. Voltage control on induction machines must be done with capacitors supplying reactive power.

Figure 5.11 shows such a system. At no-load, the IMAG is only connected to a set of excitation capacitors which permits the establishment of (about) nominal voltage at nominal frequency. With increasing load, voltage drops and additional capacitors are connected step by step to re-establish voltage by supplying additional reactive power to the IMAG. The switches are controlled by electronic voltage relays which are actuated at a certain minimum and maximum voltage value (relays with a hysteresis of 10% which gives a voltage variation of ±5% of nominal voltage). The relays must be logically interconnected; i.e any additional relay is not under power until the preceding capacitors are actuated and, when power drops, remains locked until the higher-level capacitors are disconnected. Note that the load for which a certain capacitor will be switched on depends on the frequency. In Figure 5.11, frequency has been kept constant; in reality, frequency may fluctuate due to the hand-regulation of the turbine opening.

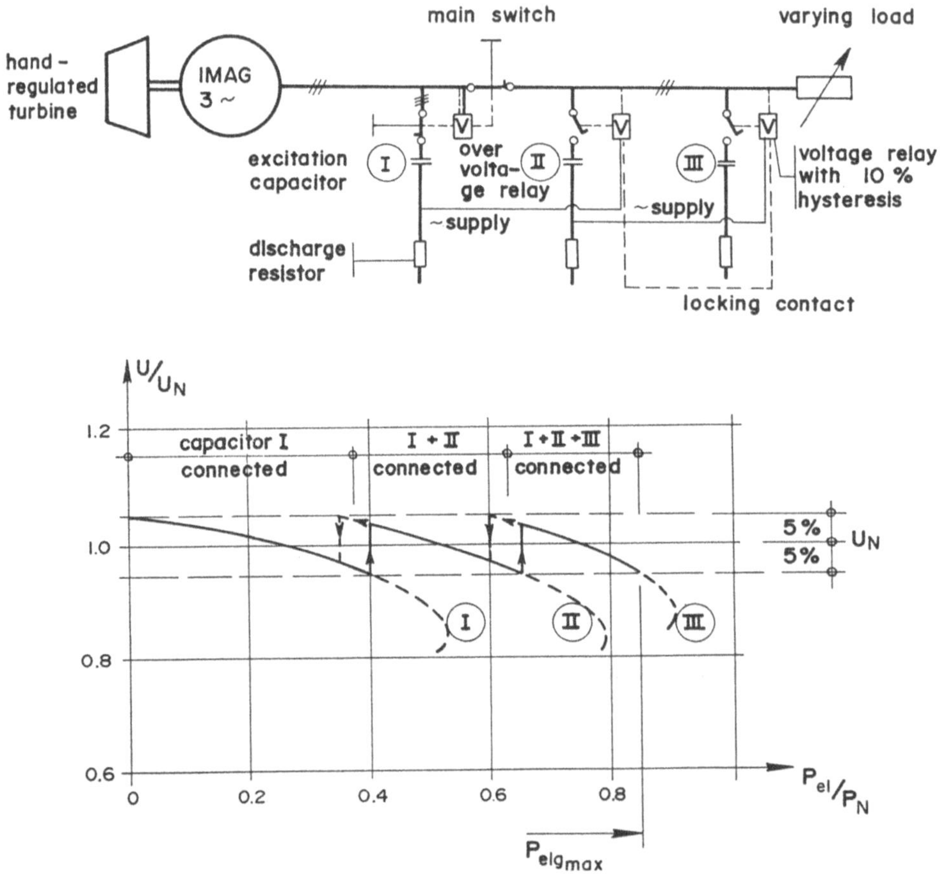

FIGURE 5.11 :
Schematic layout of a hand-regulated plant with stepped capacitors for voltage control and graph of voltage fluctuations for a varying load

Sizing Capacitors

For appropriate sizing of capacitors, the development of voltages as indicated in Figure 5.11 should be known. As a first approximation, the excitation capacitors for no-load could be taken as slightly above full compensation of the motor reactive power. Two further steps are usually sufficient if a voltage variation of ±10% is permissible; the value of the intermediate set of capacitors may be taken as the mean of the no-load and full-load set. Adjustment of the capacitor size on site is to be foreseen in any case.

5.3.4 IMAG in Conjunction with a Load Controller

Induction generator controllers (IGC) have been developed recently both in the UK and in Switzerland. The generating unit continuously operates at constant output. Any excess power not absorbed by the consumers is diverted by the IGC into a ballast load (water or air heaters = heat sink). It works on the principle that the IMAG will speed up and produce overvoltage if consumed power is lower than generated power (see also Figure 5.4 above). Therefore, the IGC senses voltage to assess how much active power must be dissipated in the ballast load. Hence, the IGC is essentially a speed/frequency controller which <u>**controls active power**</u> (through voltage regulation) on the consumer side rather than on the turbine (such as the flow control of conventional speed governors).

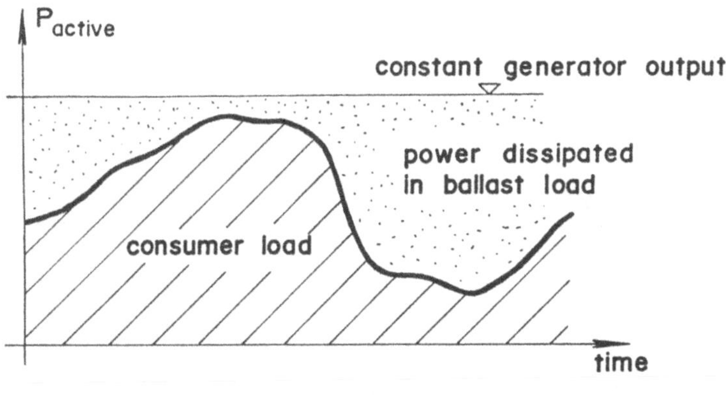

FIGURE 5.12 :
Working principle of a induction generator controller (Active power control)

<u>**Reactive power control**</u>
As long as the electrical loads on the IMAG are purely resistive, the reactive power demand of the IMAG remains constant and can be supplied by a fixed size of capacitors (calculation see above). When an inductive load (e.g., an induction motor) is connected to the IMAG, the reactive power demand of the system will increase. Note that this additional reactive power Q is required by the inductive load (external or load power factor $\cos\phi < 1$) and is not part of the magnetizing requirements of the IMAG; these remain the same due the constant active power demand (P_{load} + $P_{ballast}$ always constant, see Figure 5.12 above). However, the fixed set of capacitors cannot supply additional reactive power without a corresponding drop in voltage (see Figure 5.13a below). The IGC senses this voltage drop and subsequently reduces the power dissipated in the ballast load. Being basically an active power controller, the IGC can only compensate for the active power of the inductive load. The reactive part can be supplied only by an increase of speed / frequency of the IMAG (negative slip increases). Figure 5.13b shows this behaviour. The increase of frequency has two effects: firstly, the magnetizing requirements of the IMAG are reduced (magnetization curve steeper) and, secondly, the Var produced by the capacitors increase. Since these two effects both provide for the reactive power demand of the load (at nominal voltage), the increase of frequency is fairly small, typically 5% for a load power factor of 0.9 and 10 % for 0.8 (source ITDG). This is mainly thanks to the fact that modern IMAGs generally show a high level of saturation (non-linearity of the magnetizing curve starts at low voltage, see motor B on Figure 4.3b above) which can produce large changes in reactive power at low frequency variations. Hence, from the point of view of stability (frequency variations due to inductive loads), the saturated IMAG has an advantage over the non-saturated one. This is quite the contrary to what is required from the point of view of efficiency and electric power output in generator mode (see selection parameters, Chapter 4).

Nevertheless, we recommend that motors with a low level of saturation be preferred since frequency variations of -0% and up to +10% are usually acceptable for rural applications. Additionally, improvements are possible by direct power factor correction on individual loads which are highly inductive.

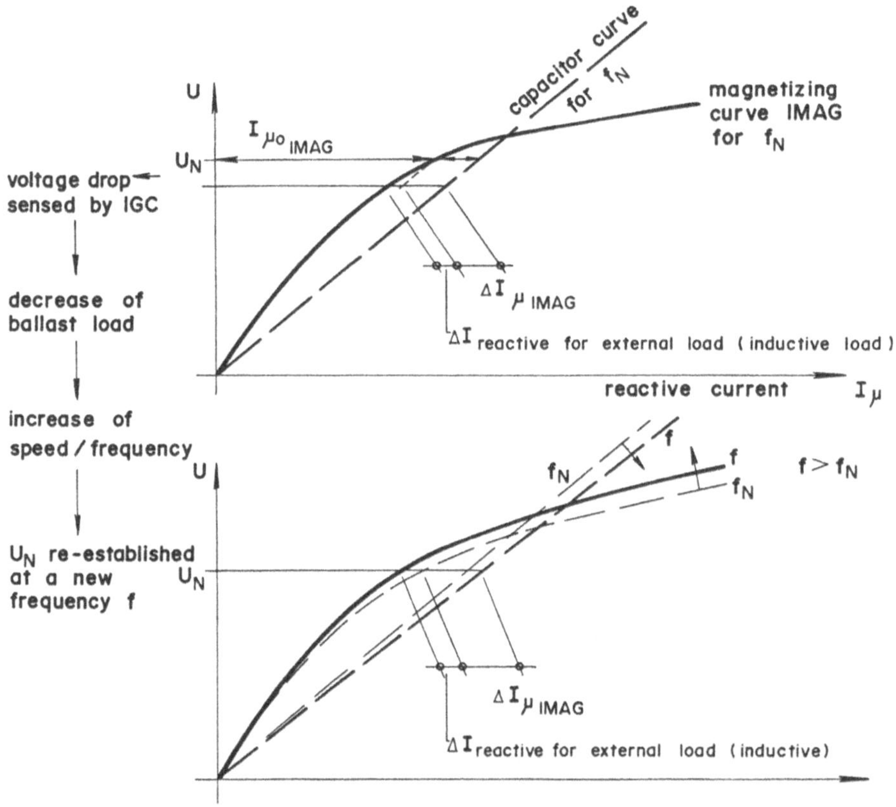

FIGURE 5.13 :
Frequency variation of an IGC-controlled IMAG when loaded with an inductive load

A problem arises if large induction motors are started direct-on-line in an MHP using an IMAG. The reactive power demand at start-up may reach 6 to 8 times that of normal operation (high starting current at low power factor, see circle diagram, slip = 1). Compensation of such high reactive power requirements by capacitors and their automatic disconnection when the motor comes to speed (e.g with a DC relay and a delay time capacitor) is not economical for larger motors. Direct-on-line start of large motor loads may therefore cause very high frequency variations of the IMAG (danger for other consumers) and may even cause excitation to collapse.

Generally, induction motors of rated input power larger than 20 to 25 % of the IMAG electrical output cannot be started (direct-on-line) in an isolated system using a load controller. In this case, it might be better to use a synchronous generator rather than an IMAG.

Figure 5.14 provides a schematic circuit diagram of an IGC. Generated voltage is measured and compared with a reference value (nominal voltage). As soon as there is a voltage difference, the comparator/amplifier sends a signal (corresponding to the detected voltage difference) to a converter where the signal is modulated to rectangular waveform. The pulse width of this signal is proportional to the voltage difference and is a measure for the amount of power to be dissipated in the ballast load. The modulated signal supplies a transistor (electronic switch) which opens and closes according to the rate of the signal, i.e. the ballast load circuit stays open only during the on-time of the modulated signal. Since a transistor allows current to flow only in one direction, the AC output of the IMAG must be rectified by a full-wave bridge rectifier comprising four diodes.

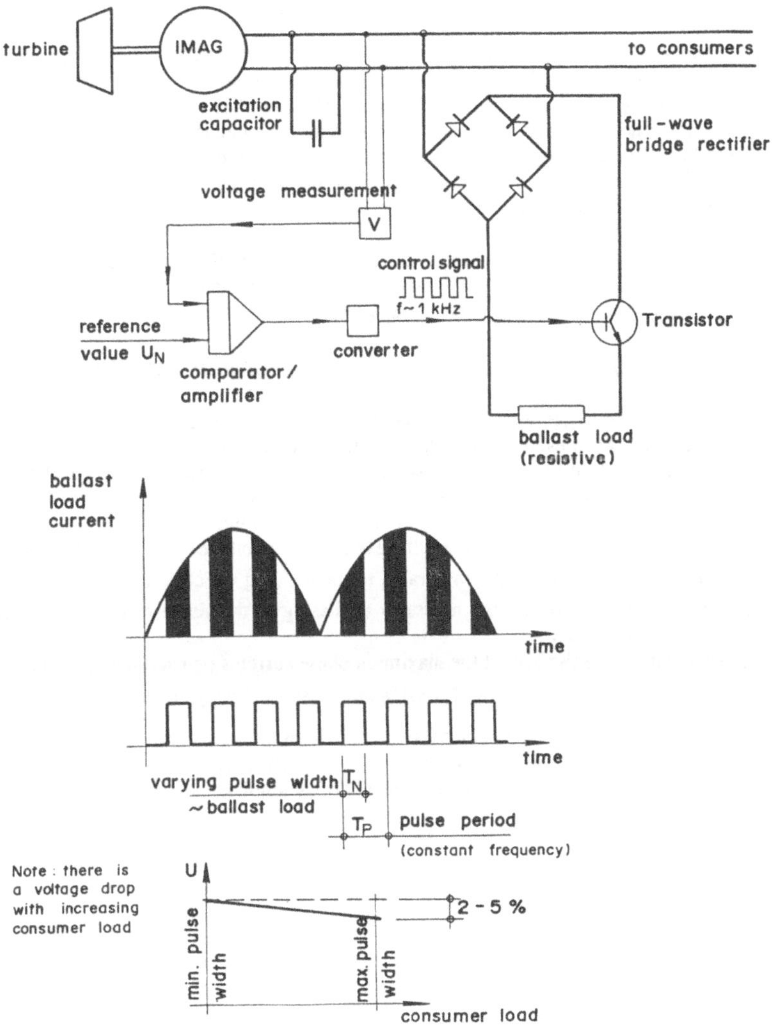

FIGURE 5.14 :
Simplified circuit diagram of an induction generator controller (IGC)

A list of manufacturers of IGCs is given in Appendix F.

5.4 Three-Phase Motor as Single-Phase Generator

5.4.1 General

For very small electrification schemes (<10kW), single-phase electricity generation and distribution systems might have advantages over three-phase systems. For small three-phase schemes, individual single-phase loads might become a significant percentage of the total generator output and balancing load on the generator at any instant of operation would be extremely difficult. Thus, the main advantage of a single-phase system is that the problem of balancing of loads does not exist.
Additionally, costs might be lower for a single-phase system because switchgear, monitoring, control and protection equipment are needed for only one rather than for three phases. For larger schemes (P>10kW) with longer distribution lines, this advantage is usually ruled out by the higher costs of the conductor material for single-phase networks which have to carry total generator power on a single circuit (including two wires) whereas the wires of a three-phase system carry only a third.

The use of an induction motor as generator for a single-phase system can be a very economical solution as much as for a three-phase system. Despite the fact that single phase motors in the small power range (up to 3 - 7.5 kW) exist, we nevertheless recommend the use of three-phase IMAGs. Single-phase motors are equipped with starting devices (capacitors and the corresponding switches to short them out after start up) which may not be relevant for generator operation. Some of these devices must be disconnected for generator operation and are therefore a wasted investment. Even without these devices, single-phase motors are usually more expensive than three-phase ones.

Technically, single-phase motors can be used in the same way as the three-phase ones; i.e capacitors must be connected to the terminal to provide excitation. The size of the capacitors can be determined from the generator $\cos\phi$ as shown for three-phase IMAGs above.

5.4.2 Balancing the single-phase load over the three phases of the IMAG - the C-2C connection

When connecting a single phase load on a three-phase IMAG, each phase will be loaded differently. Current and voltage will not be evenly distributed over the windings and the phase in parallel with the load will be overloaded long before the other two windings reach their current limits; i.e. the IMAG cannot supply the same power as with a balanced three-phase load. Connecting excitation capacitors in a so-called C-2C mode (see Figure 5.15 below) considerably improves the situation.
The C-2C connection proposes a scheme where the excitation capacitance C_{phase} is connected in parallel to the load and a capacitance of twice C_{phase} in parallel to the winding which follows the load terminal in the sense of rotation. This latter provides the necessary dephasing of the single-phase load so that it appears balanced for the three-phase IMAG.
However, perfect balancing of the load at the maximum phase currents (= maximum permissible load as on a three-phase grid without overheating the IMAG) is only possible for a generator power factor $\cos\phi_g = 0.5$ or $\phi = 60°$ (which is a standard value for IMAGs of rated motor power around 2 kW, see Chapter 4 above). Figure 5.15 below shows the voltage and current phasor diagrams for this power factor. The excitation capacitance C_{phase} for the C-2C connection can be calculated as follows:

valid for $\cos\phi_g = 0.5$:

$$C_{phase} = \frac{1}{\sqrt{3}\, 2\pi f\, R_{load}} \tag{5.6}$$

$$\text{where } R_{load} = \frac{U}{I_R} = \frac{U}{1.5\, I_{phase}} \tag{5.7}$$

For any other $\cos\phi_g$, a single-phase load can only appear balanced for the 3-phase IMAG if resistances or inductances in addition to capacitors are connected to the windings of the IMAG. However, this is a fairly complicated procedure and is seldom applied for IMAGs. It is usually more economical to apply the C-2C connection for any power factors $\cos\phi_g$ but to accept a slight imbalance of the phase currents and voltages. In this case, a derating of the maximum generator power output is required in order not to exceed the thermal limit of the machine. Note that balanced conditions could also be attained for generator power factors other than 0.5 by allowing some variations of frequency and/or voltage (see also section 5.5 below).

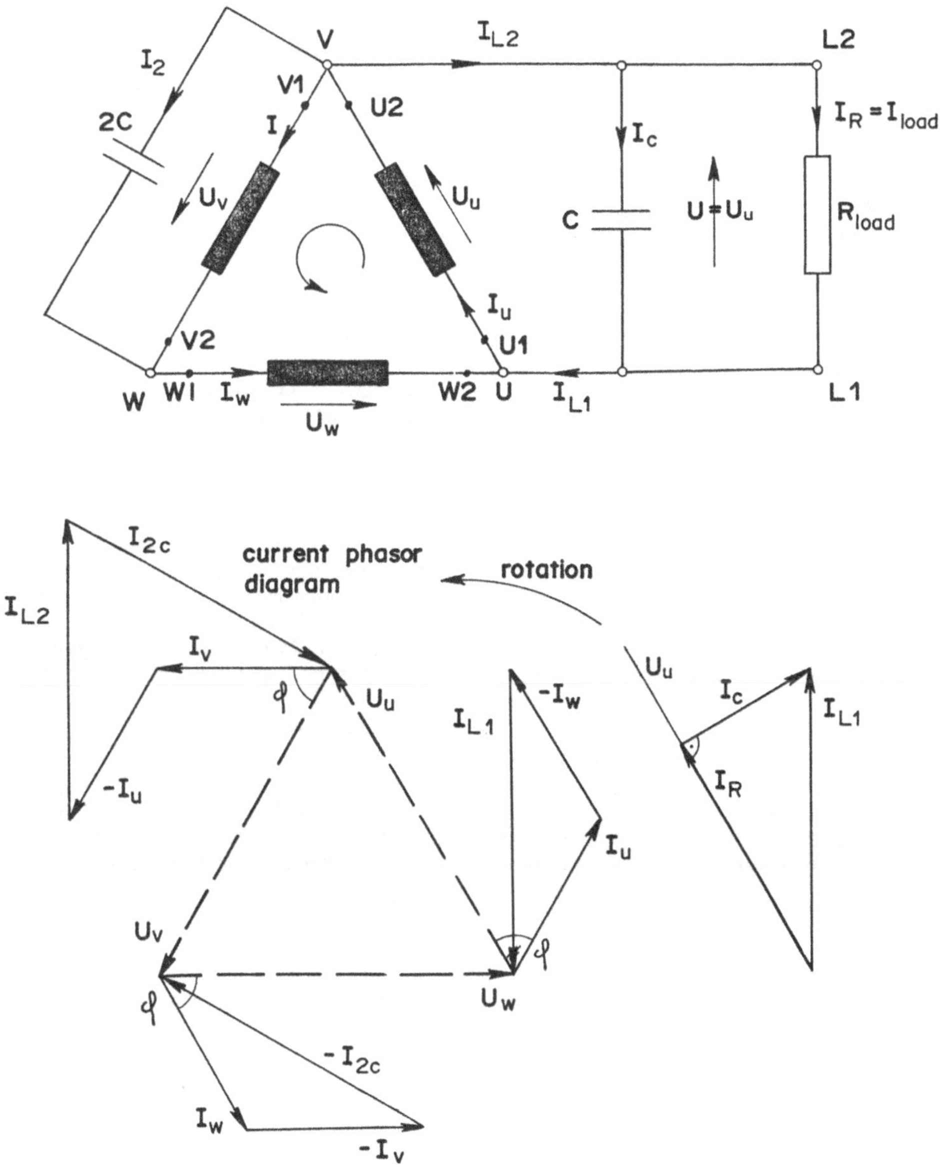

<u>balanced phase voltages and currents if $\cos\phi_g = 0.5$</u>

- from phasor diagrams: $I_R = 1.5\, I_U = 1.5\, I_{phase}$

FIGURE 5.15 :
C-2C connection for a three-phase IMAG loaded with a single phase resistive load, and voltage and current phasor diagrams for a $\cos\phi_g = 0.5$.

The maximum electrical output of the three-phase IMAG connected to a single-phase system may be estimated as follows:

$$\boxed{P_{el_g} \max = \frac{S}{2}} \qquad (5.8)$$

where S = apparent power = $\sqrt{3}\, U_{line}\, I_{line}$ with U and I from motor name plate

5.5 What to do if the expected power output of the selected IMAG is not attained

The selection procedure of an IMAG is always subjected to a certain inaccuracy; this is mainly due to the fact that selection diagrams presented in this manual cannot cover the large number of different motor designs available on the market. It might well be that the selected IMAG will not deliver the expected power on commissioning. The following sections provide some simple measures to partly **improve a poor performance of an IMAG**.

5.5.1 Parallel Operation

Corrective measures for an IMAG operating in parallel to a large network are limited. The operating conditions are determined by the grid (voltage and frequency) and by the prime mover (mechanical input from turbine). If the desired electrical output is not reached within the thermal limit of the IMAG (see Chapter 4), only the exchange of the IMAG can actually improve the situation. It is therefore advisable to select IMAGs of a slightly higher rating than absolutely necessary from calculations (= **overrating**). The drawback of the overrated IMAG is that the efficiency at part-load operation will be lower than if the turbine and IMAG ratings are closely matched.
A possible remedy appears to be the **change from delta connection at full mechanical power input to star connection at lower turbine input** (low stream flows during the dry season). Both efficiency and power factor can thus be improved. Note that phase voltage decreases by square root 3 when changing from Δ to Y connection and phase currents will increase. The change can only take place at about 50 to 80% of maximum IMAG output in order not to exceed the rated phase current of the machine. Figure 5.17 below presents this method for a 1.1 kW motor used as generator and the possible increase of power generated.

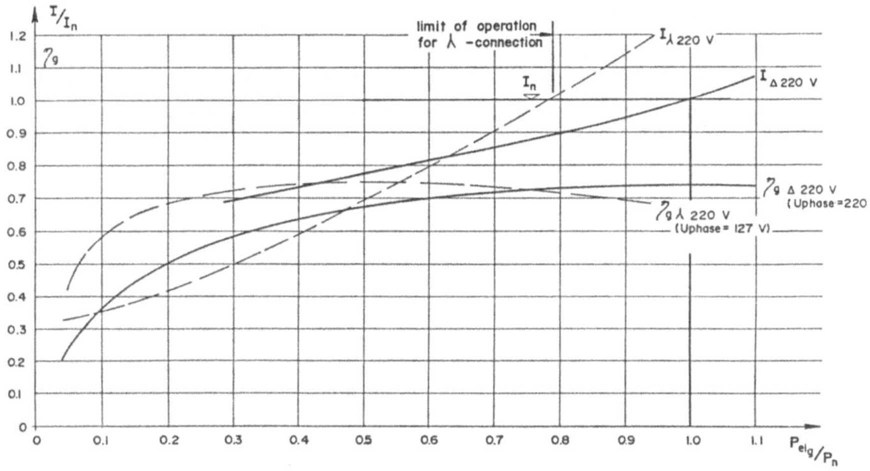

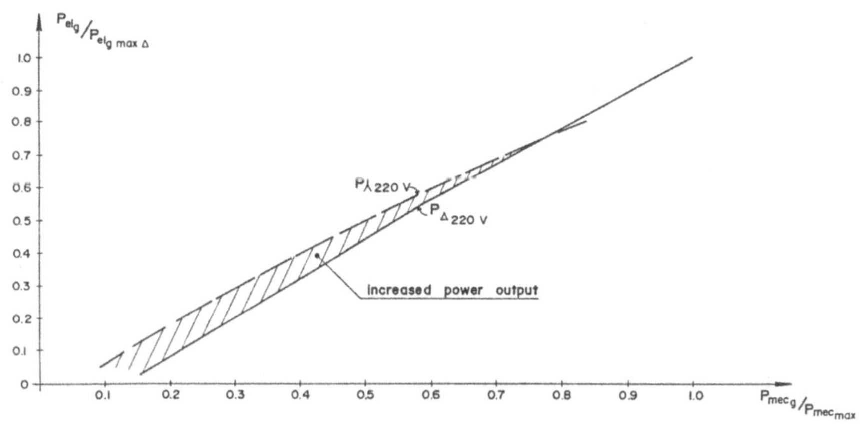

FIGURE 5.16 :
Improvement of efficiency and power factor at low power input by changing from Δ to Y connection (for IMAGs in parallel operation)

5.5.2 Stand-alone Electricity Generation

In isolated operation, voltage and frequency of the generated electricity are determined by the generating unit and may be varied over a certain range if necessary. It has been found that by **allowing a slightly higher frequency and a lower voltage** than according to standard values, an improved electrical output is possible without exceeding rated currents of the machine or its thermal limit.
This can be a convenient remedy if the desired IMAG output is not reached, for example due to a saturated machine. Figure 5.18 below shows the possible increase of electrical power by reducing the reactive part of the total current, i.e I_{total} is maintained at $I_{nominal}$ by reducing the value of the excitation capacitors. The consequence of this is of course an **increase of frequency** (brought about by an increase of turbine speed/flow); but as the diagram shows, frequency variations are low in comparison to the increase of

electrical power to be gained. A frequency variation of +10 % (= 55 Hz instead of 50 Hz) is usually acceptable for rural applications. As a rough estimate we can assume that the power increase is proportional to the square of the frequency variation; thus, for an increase of frequency of 10% we arrive at a maximum electrical power of 1.21 * P_{el_g} at nominal frequency. This increase is usually worthwhile to be considered. However, the improvement is only possible if the required increase of flow is available.

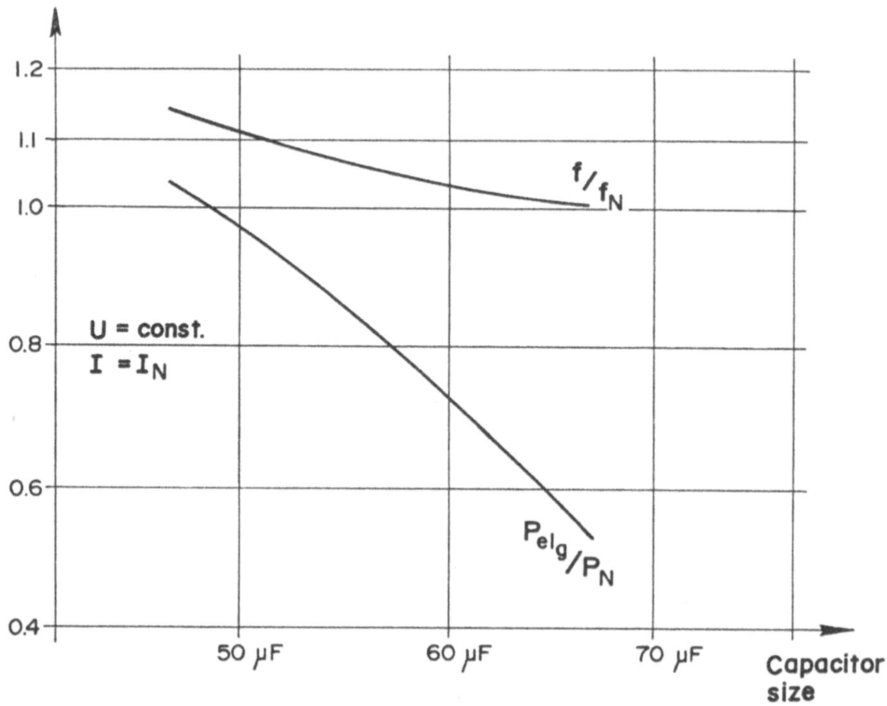

FIGURE 5.17 :
Variation of electrical power output in function of excitation capacitance by allowing frequency to vary ($I = I_n$, U = constant). From tests on a 2.2 kW motor.

Similarly, electrical power output can be increased by <u>allowing voltage to drop below the nominal value</u>; i.e., capacitance is reduced as above, but voltage is not entirely brought back to nominal through adjustments of the turbine opening. As Figure 5.19 shows, the increase of power to be obtained depends very much on the saturation level of the machine. IMAG 3 (2.2 kW, normal saturation level) shows an increase of about 10 % in electrical power upon a voltage drop of 5 % (e.g. 380 to 360 V) while the output of IMAG 2 (P_n = 2.2 kW, highly saturated with almost no output in generator operation at 380 V) can be brought to an acceptable level (P_{el_g} /P_n = 0.7) simply by allowing the same voltage drop of 5 %.

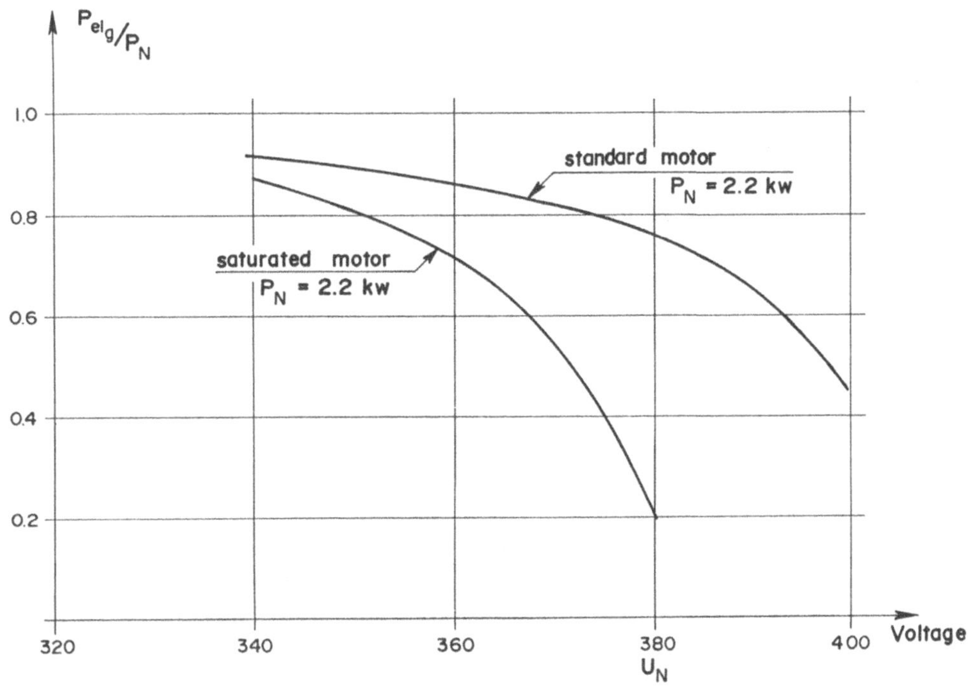

FIGURE 5.18 :
Variation of power output in function of varying voltage of two IMAGs of 2.2 kW rated motor power but with different saturation levels

5.6 Overview of monitoring equipment and protection required in MHPs using an IMAG

Figure 5.20 shows the principle layout of an MHP using an IMAG; the monitoring and protection equipment is divided into four groups:

turbine control and protection, generator protection, monitoring equipment and grid protection. Depending on the configuration of the plant (stand-alone or parallel, type of governor/controller) some of the equipment listed is not necessary or does not apply. Table 5.21 gives an overview of recommended equipment for MHPs in conjunction with an IMAG.

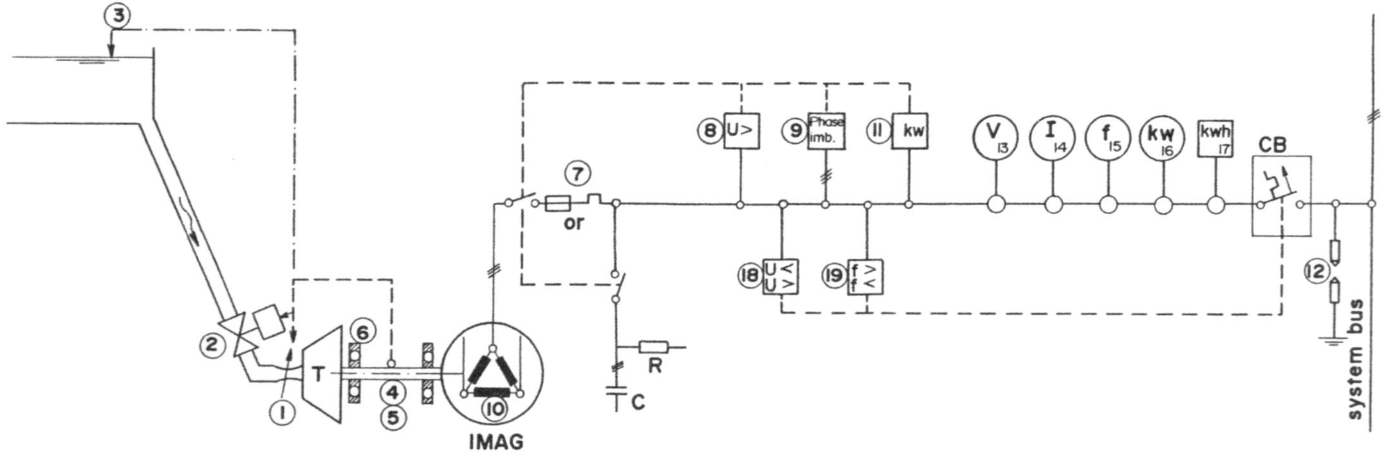

Turbine Control and Protection	Generator Protection	Station Monitoring	Grid Protection
1 Turbine gate (guide vane)	7 Overload / short-circuit	13 Voltmeter	18 Over- / undervoltage
2 Emergency shutdown (valve, jet deflector)	8 Overvoltage	14 Ammmeter	19 Over- / underfrequency
3 Water level control	9 Phase imbalance	15 Frequency-meter	
4 Tachometer	10 Winding temperature	16 Wattmeter	
5 Overspeed trip	11 Reverse-power relay	17 kWh-meter	
6 Bearing temperature	12 Lightning arrestor		

FIGURE 5.19 :
General layout of an MHP using an IMAG: Monitoring, control and protection equipment

TABLE 5.20 :
Recommendations for monitoring, control and protection equipment for IMAGs in different configurations

Type of Equipment	Parallel Operation	Stand-alone Electricity Generation			
		Unregulated	Stepped capacitors	IGC	C-2C
Turbine Protection and Control					
1) Turbine gate	O f(water level)		O (handregulated)	(o)	
2) Emergency shutdown	O			(o)	
3) Water level control	O				
4) Tachometer	(o) or frequency meter				
5) Overspeed trip	(o) or overfrequency trip			(o)	
6) Bearing temperature	(o) not below 50 kW			(o) not below 50 kW	
Generator Protection					
7) Overload / Short-circuit	O	O	O	O	O
8) Overvoltage	O	O	O	O	O
9) Phase imbalance	(o)			(o)	
10) Winding temperature	(o) not below 50 kW			(o) not below 50 kW	
11) Reverse power relay	O				
12) Lightning Protection	O (waterlevel telecontrol)			O (load controller protection)	
Monitoring					
13) Voltmeter	O	O	O	O	O
14) Ammeter	O	O	O	O	O
15) Frequency - meter	(o) or tacho				
16) Wattmeter	(o)				
17) kWh-counter	O	(o)	(o)	(o)	(o)
Grid protection					
18) Over- / under- voltage	O				
19) Over- / Under-frequency	(o) or overspeed trip				

6. SAFETY AND PROTECTION

6.1 General

The aim of using induction motors as generators is primarily to reduce the costs of electricity generation in micro-hydropower plants. Often in small MHPs, the cost-reduction approach is extended into many other components of the scheme such as turbine manufacture, governing and also electrical system protection.
It may be due to ignorance or because of applying cost-reduction principles too literally that safety and protection in small electrification schemes are sometimes criminally neglected. The dangers of electricity remain essentially the same whether we deal with 0.5 kW or with 500 kW installations and efforts should be taken to protect both people and equipment.
The following chapter provides some basic principles of safety and protection in small-scale electricity generation and distribution systems. It is intended for the non-specialist in this field who, in any case, will have to contact the experienced electrical engineer and fitter for more detailed planning and implementation.

6.2 The Danger of Electricity

6.2.1 To People

An electrical current flowing through human beings can knock the person unconscious, cause serious burns, stop the person's breath or stop his heart beat. The danger of being seriously injured or killed by an electrical current depends on:

- ***the nature of the current:*** both AC and DC currents can be dangerous but AC of the commonly used frequencies of 50 or 60 Hz is the most dangerous current, especially as far as provoking a standstill of the heart is concerned; high frequencies (> 500 kHz) do not present any danger.

- ***the path the current takes across the human body:*** a standstill of the heart can be caused if the current of sufficient intensity passes through the upper part of the body.

- ***the intensity of the current:*** the current flowing through a person depends on the voltage drop and the resistance of the human body (esp. its humidity). **Electrical systems of voltages below 50 V (AC or DC) do not produce currents of dangerous levels.**

Intensity of the current	*Effects on the human body*
< 15 mA	an electric shock can be felt; above 15 mA the victim cannot free himself from the current carrying wire or casing (let-go current)
20 - 30 mA	the muscles are contracted heavily and trouble in breathing can cause death after about one minute
80 mA	the heart contracts superficially and blood circulation is stopped
> 3 A	- severe burns - intoxication of the body and death after a few days

6.2.2 To Installations

Fault currents and short-circuits can destroy the electrical equipment and, usually more severe, can cause fire in buildings if no measures are taken to detect abnormal situations and to disconnect any defective circuit immediately.

6.3 Responsibilities in case of electrical accidents and damage

When establishing an electricity generating scheme, client, designers and contractors should clarify the responsibilities of the different parties involved in case of accidents and damage caused by faulty or defective installations. Rules and regulations concerning these may slightly differ from country to country. We will give hereafter some of the legal guide-lines applied in continental Europe.

According to the Swiss Federal Law, it is the owner / operator of an electrical installation who is made responsible in the first place for the injury or death of people or any other damage caused by an electrical defect on his premises. Since the owner / operator is seldom an electrical engineer himself, he cannot judge whether the work of the fitter or engineer he had engaged corresponds to the standards. The government therefore authorizes only experienced electricians with an approved education and training to carry out any installation, modification or repair of electrical systems.
The implementation of electrification schemes by other than qualified electricians, be it for cost-reduction purposes or any other reason, is therefore an offence, at least in the sense of European laws. But the responsible micro-hydropower engineer, aware of the dangers of electricity, would do wise to apply the same principle in any other country.

6.4 Fault Conditions and Protective Measures

6.4.1 Protection of People - System Earthing

Either of two principles may be applied to protect people from being electrocuted:
- the system voltage is at any time below 50 V
- if voltage is above 50 V, the corresponding fault current causing electric shock (or fire) must be cut immediately

The latter principle is the generally applied solution for village electrification systems and comprises earthing of metallic frames of electric appliances. Figure 6.1 shows how the system works in case of an electric fault.
The appliance has an insulation defect in the line L1 and the metallic frame comes under voltage. Anyone getting in contact with this frame will be electrocuted because the human body will close the fault current path through the earth and the grounded star point of the generator or transformer (see Figure 6.1a). The fuse in line L1 might not trip because the resistance of the fault current path including the human body can prevent a current sufficiently high to melt the fuse.
An improvement of the situation can be brought about if the metallic frame of the appliance is grounded through an earth electrode (metal plate buried in the ground; depth depends on the nature of the soil; humid soil near river preferable). This system is called protective earthing (or direct earthing, TT-system; see Figures 6.1b and 6.2). When an insulation defect occurs, the earth electrode will provide the path for the fault current to return to earth and the individual touching the appliance will not be electrocuted.

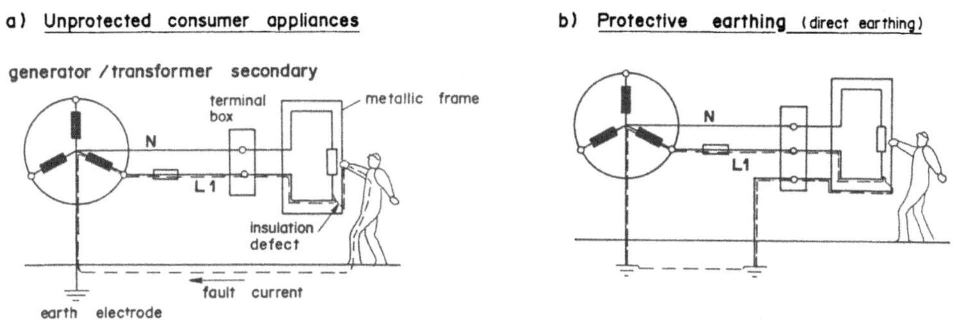

FIGURE 6.1 :
a) Fault current through human body in the case of an insulation defect on an unprotected appliance
b) Fault current path with protective earthing (TT-system)

The individual protective earthing of appliances has the disadvantage that the resistance of the earth is seldom sufficiently low to provoke a short-circuit in the line with the fault and a subsequent tripping of the fuse. In supply systems with protective earthing it is possible that dangerous fault currents may persist for an unlimited time. People close to the earth electrodes can be endangered due to the voltage drop between their two feet (step voltage). The protective earthing of individual appliances is, despite its simplicity, not a very safe and effective system unless the required earth resistance can be guaranteed (the resistance of the earth may vary with time and has to be checked in accordance with regulations).

The use of a conductor as a return path for a fault current (instead of the earth) is a widely applied system; this can either be accomplished by a separate conductor or the neutral wire. Figure 6.2 shows various earthing arrangements which conform to international practice. For clarity, only one phase of the supply to a customer's installation is shown. The codings used correspond to international standards:
T = earthing (Terre)
N = Neutral point of the power system (generator or transformer star point/neutral)
S = Separate conductors for earthing and neutral
C = Combined conductors for earthing and neutral, the so-called PEN conductor
PE = Protective Earthing

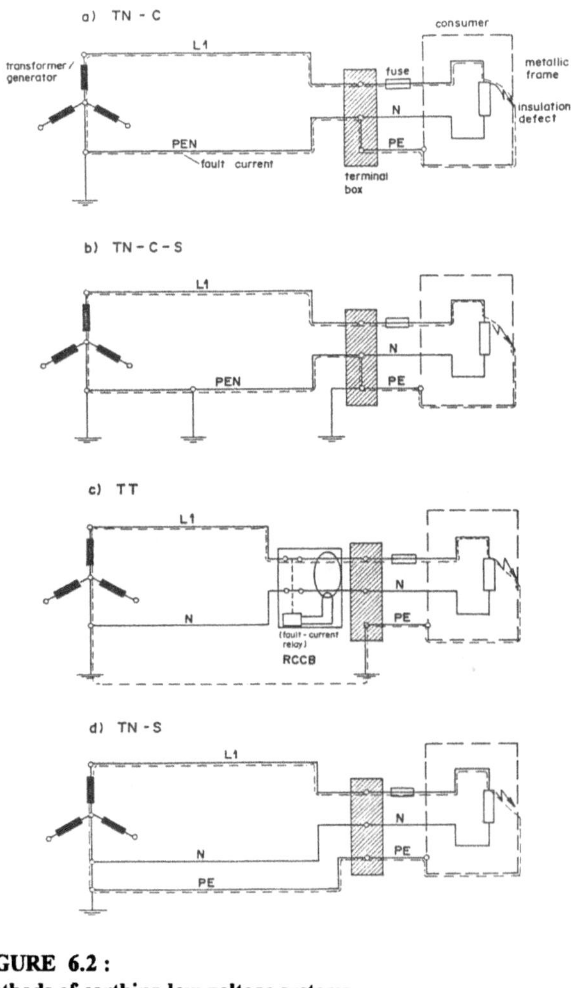

FIGURE 6.2 :
Methods of earthing low-voltage systems
a) TN-C system: common neutral and PE conductor
b) TN-C-S system: same as a) but with multiple earths
c) TT system: electrically independent neutral and protective earthing with RCCB
d) TN-S system: separate neutral and protective earth conductors

a) TN-C system: The metallic frames of appliances are connected to the neutral wire which is therefore used in two ways: during normal operation as a life wire (return path for single phase appliances or for unbalanced 3 -phase systems) and in case of fault condition as a protective conductor (return path for the fault current to trip the fuse). Note that the neutral itself must never be fused since a broken fuse would interrupt not only the supply but also the protective system; and even worse, it would put all metallic frames beyond the broken PEN conductor under voltage even without a fault condition (see Figure below)

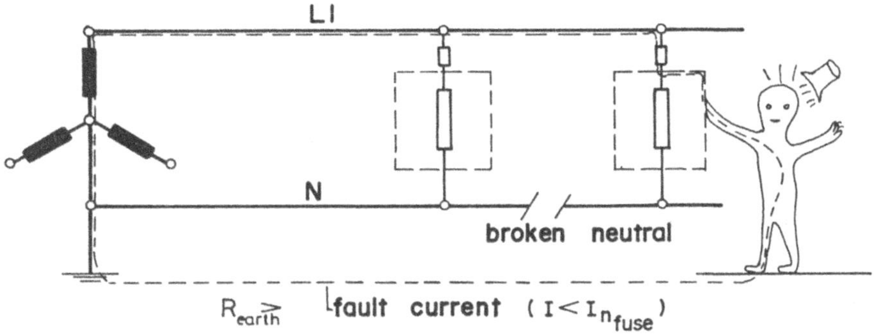

b) TN-C-S system: The neutral is earthed at a number of points. This ensures that in case of a broken neutral wire, the section beyond the broken PEN-conductor will still have an earthing path.

c) TT arrangement: The drawback of the protective earthing arrangement (individual earthing of customer's appliances, see Figure 6.1 above) can be avoided if a suitable device, a so-called residual-current circuit breaker (RCCB or fault current relay) is installed at the entrance to the customer's premises. When a fault condition occurs, this device senses the current difference (leakage current) between the two life wires and opens the circuit. RCCBs are also available for three-phase systems.

d) TN-S arrangement: The protective earth conductor PE is separated from the neutral wire. Though this arrangement requires the largest amount of conductor material, it is a relatively safe system and accidents are rare.

The crucial point with all the protective earthing systems described above is the fact that mixing up phase and PE(N) conductor in the terminal box puts the metallic frame of the electrical appliances under voltage even without an insulation fault. Wiring of installations should therefore only be undertaken by an experienced fitter.

6.4.2 Protection of the Electrical Installation

Apart from an insulation defect as described above, other fault conditions can occur in an electric system which may not immediately put people in danger but which can cause severe damage to the installation. System faults can be the following:

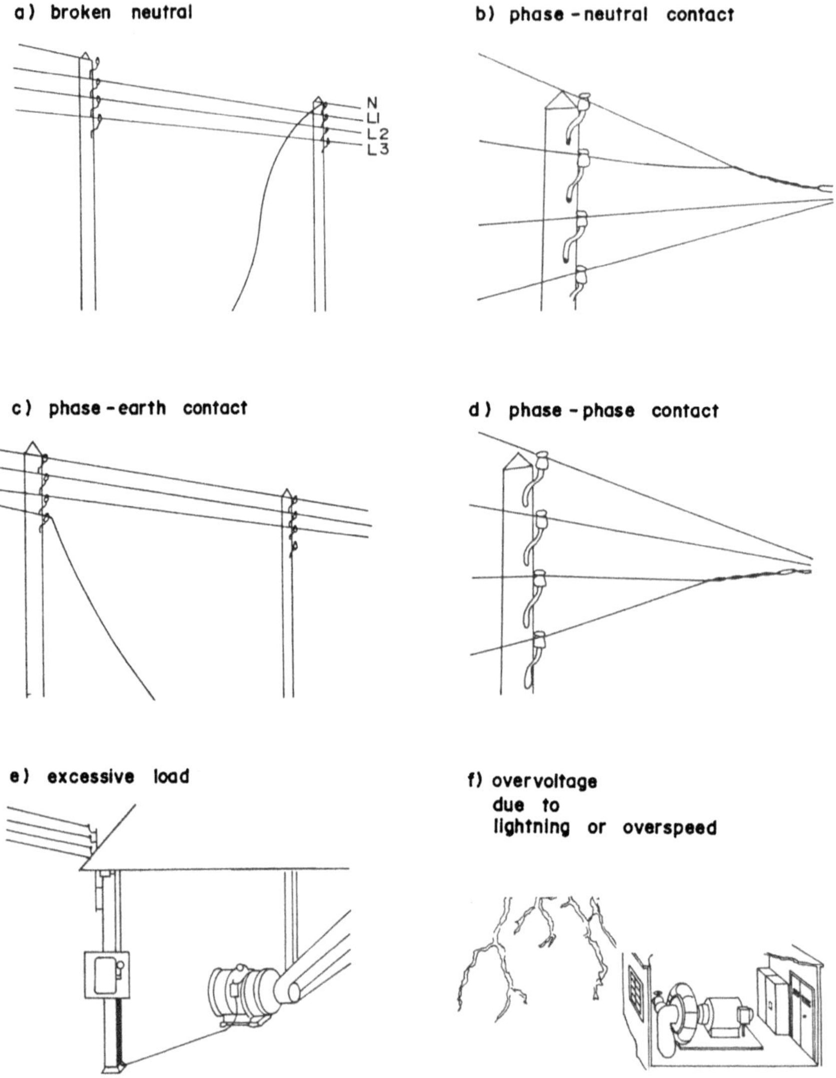

FIGURE 6.3 :
Examples of system faults

Short-circuit currents, overcurrents and overvoltage as shown in Figure 6.3 will cause conductors and windings to heat up, burn insulations and may finally provoke fire in installations and buildings. At the occurrence of abnormal conditions, the electrical circuit must be opened in a minimum time to minimize damage. Various kinds of equipment are used to sense the fault condition, to break a current and to reclose the circuit once the fault has been removed.

6.5. Equipment

6.5.1 Switchgear

Switchgear is a general term covering electric switching devices and their assemblies of various capabilities:

- A ***circuit breaker*** (CB) is a switch designed to open even when a high current is flowing. When a protection device (see section 6.5.2) senses a fault current, it sends a signal to the circuit breaker to open the contacts. Some small circuit breakers (miniature circuit breakers, MCB) or overcurrent relays direct on the line open by themselves when they are forced to handle excess current. Breakers can also be opened manually.

  High currents will draw an arc between contacts, if the breaker opens. This arc must be extinguished as quickly as possible; air or oil may be used in circuit breakers as extinguishing medium. In a vacuum circuit breaker (VCB) the contacts open and close within a small vacuum bottle and no arc can be conducted. For low-voltage systems, air circuit breakers are usually sufficient. A circuit breaker is designed to operate (break) many times between maintenance periods.

- A ***switch*** is a device capable of making, carrying and breaking currents under normal operating conditions. Under fault current conditions, the contacts may melt together; the switch is therefore not designed to make or break fault current.

- A ***disconnector*** is defined as a mechanical switching device which enables off-load breaks to be made on circuits of medium or high voltage lines. The contactors therefore provide in the open position a specified isolating distance. A disconnector is not capable of making or breaking currents.

6.5.2 Overcurrent protection devices

Basically, there are two types of overcurrents to be distinguished:

Firstly, the overcurrent of high intensity, i.e., a short-circuit current, where the protection device must open the circuit very rapidly to avoid damage,

and secondly, the overcurrent which exceeds the nominal current by only a small amount which is basically an overload situation. Overload of short duration is not dangerous (may occur at motor starts etc.) but if it persists for a long time, conductors and installations will be heated up gradually. The overload protection device must therefore cut the circuit only after a certain time delay.

- **Short-circuit protection: Fuses**

  Fuses act as both a protective and a disconnecting device and were the earliest such devices used on electrical systems. They basically consist of a metallic element which melts and becomes discontinuous at a relatively high current, thus preventing the further passage of the current. The disadvantage of fuses is that, having melted upon a transient overcurrent, they require replacement before the protected circuit can carry load again. Dependent on the intensity of the fault current, the operating time of a fuse can take from a few tenths of seconds to hours. Figure 6.4 shows a typical time/current characteristic of a fuse.

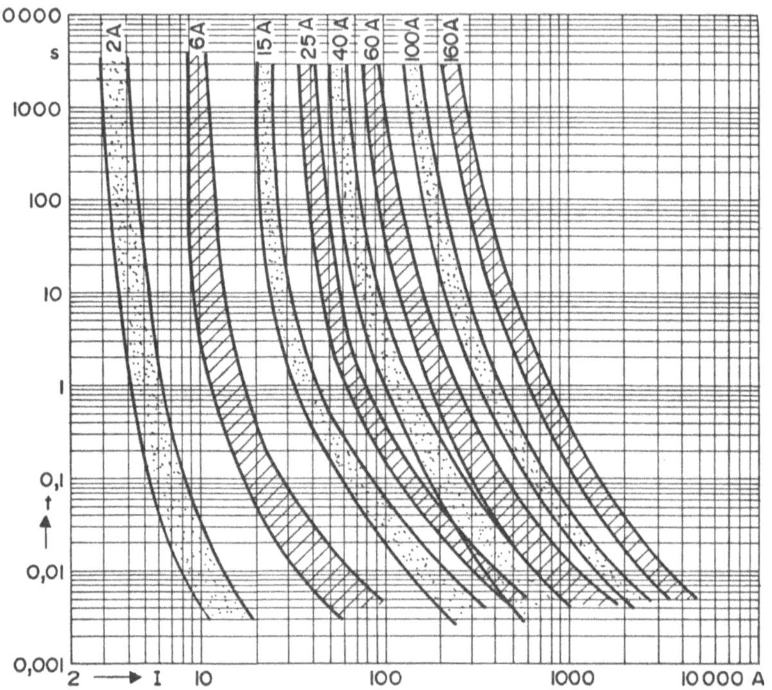

FIGURE 6.4 :
Characteristics of fuses

Due to the fact that fuses need replacement after a single tripping operation (stock of fuses), they are generally used at the highest protection level, i.e they trip only if all other protection devices fail (back-up protection).

- **Overload Protection - Overcurrent Relays**

An overload condition on the (synchronous) generator may occur if too many appliances are connected to the network or if a short at a customer at the end of the line has occurred without producing a current high enough to trip the fuse at the generator. (For overload conditions of IMAGs see 6.6.2 below.)

When the current exceeds a preset value, the overcurrent relay will initiate (after a time delay determined by its characteristics) the opening of the circuit (if the relay is mounted directly on the line) or, in larger power systems, to initiate tripping of the associated circuit breaker.

There are basically three different types of relays available:

- the **electromagnetic version** with a solenoid which trips instantaneously when sufficient current passes through the coil; the characteristics of this relay is shown in Figure 6.5, curve a).
- the **thermal relay** with a bimetal strip which trips after a time delay inversely dependent on the overcurrent; the characteristics of the thermal relay is therefore similar to the one of the fuse (see Figure 6.5, curve b).
- these two features may be **combined** in one relay: for an overload situation, the relay trips after a certain time delay; for a short-circuit current of several times nominal current (typically 5 to 10 I_n) the relay trips almost instantaneously (see Figure 6.5, curve c).

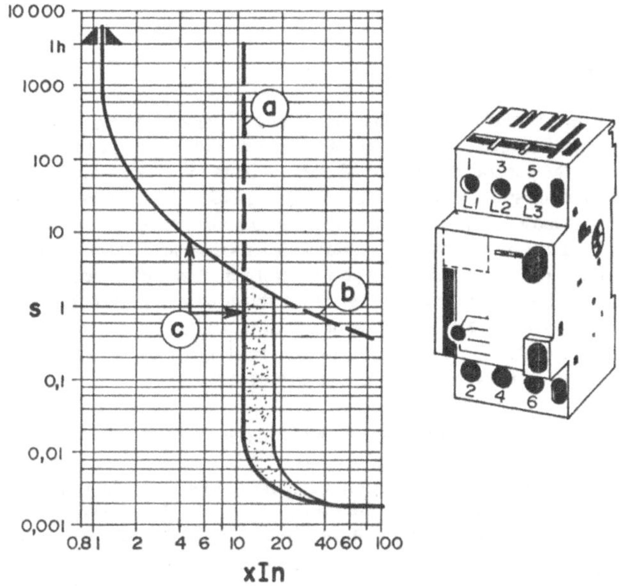

FIGURE 6.5 :
Characteristics of overcurrent relays
a) solenoid overcurrent relay with an instantaneous operation
b) thermal relay only
c) combined relay

6.5.3 Overvoltage protection devices

Overvoltage may be caused by overspeed of the generating unit (at constant excitation), failure of the voltage regulator or load controller but may also be induced by direct lightning strikes into overhead lines or to nearby ground.
Relays which disconnect any circuit if some abnormal voltage situation has arisen are usually sufficient to protect the installation from overvoltage caused by the generating unit. Note that overvoltage protection is necessary even in situations where a load controller or other voltage regulator (AVR on synchronous generators) is used, mainly as a back-up safety precaution in case of failure of the voltage controller.
Overvoltage relays are however incapable of protecting equipment and appliances from lightning strikes which can cause overvoltage surges of a few thousand volts. Distribution systems are usually protected by a so-called air-gap overvoltage protection with one metallic rod connected to a phase conductor and a second rod connected to earth. Voltages higher than the withstand voltage of the equipment insulation will cause a flashover at the air gap which then becomes an earth fault on the system similar to the one initiated by a short. Air-gaps are inexpensive but do not fully protect the system; lightning arresters using several air gaps plus resistors in series may provide a higher level of protection but are also more expensive. It is important that the air-gaps or lightning arresters are placed as close to the generator as possible, usually on the first pole outside the power-house.

While air-gaps and lightning arresters might effectively protect the electric power system, they are incapable of protecting the electronics of load controllers and telecontrolled systems (water-level control). Devices which absorb overvoltages of a dangerous level for electronics are available on the market and should be used on MHPs with load controllers or other electronics in lightning-prone areas.

Whatever sophistication of protection against lightning strikes is chosen, be it for the power or the electronic system, **protection can only be as good as the quality of the earthing**. High resistance of the earth will prevent the voltage surges from being discharged through the earth and the protection devices become ineffective.

6.6 Particularities of IMAGs

6.6.1 The problem of third-harmonic currents

A particular problem of IMAGs and induction machines in general needs to be mentioned. The non-linearity of the magnetization curve of induction machines (see Figure 3.13) causes a slight distortion of the sinusoidal waveform of the generated voltage and current. This imperfect waveform can be understood as the resultant of a number of sinusoidal waves of increasing order, i.e a series of odd harmonics with the third harmonic being the most prominent of the waves causing distortion.
Star connection of an induction machine to a grid with balanced loads should theoretically have a zero current in the neutral. In practice we will find a current in the neutral with a frequency of three times that of the fundamental, i.e the third-harmonic current. Since third-harmonic currents are caused by the non-linearity of the magnetization curve, it may be appreciated that highly saturated machines will cause more problems with third harmonics than less saturated ones. Selecting a motor with a low saturation level as proposed in Chapter 4 will therefore be advantageous also in this respect.

6.6.2 System Earthing and Internal Connection of IMAGs (Y or Δ)

The earthing arrangements as presented in section 6.4.1 above usually use the neutral point of the generator as part of the protective system. When connecting the star point of a Y-connected induction machine (motor or generator) to the neutral wire of a grid, the third-harmonic current flowing in that conductor can cause unintentional tripping of protection devices or for single-phase appliances connected between phase and neutral, the **third-harmonic current** may prevent proper operation (especially of computers and other electronics).
If the induction generator is Δ -connected, third harmonic currents circulate between phases of the machine which will in turn generate a deformation of the voltage sin-wave on the distribution system. This will generally not cause difficulties for the electrical appliances.
It is therefore advisable either to connect an induction machine in Δ or to leave the neutral point open with machines that require exclusively a star connection. For IMAGs operating in parallel to a large grid, the electricity board will usually indicate what type of connection is to be used to eliminate the negative effects of possible harmonics.

For stand-alone electricity generation (using an IMAG) the protective earthing system should make allowances for possible disturbances by the third-harmonic currents especially if sensitive single-phase appliances or protection devices are used; the following earthing arrangements may be considered:

a) **the generating unit is physically separated from the consumers by means of a transformer**; the arrangement is similar to a medium voltage (MV) transmission system using three conductors between generator and transformer only (no neutral); when using a star-connected IMAG, the neutral/star point of the machine must be isolated. (Note that the transformer must be delta-star connected; in star-star connection, distortions by third harmonics would be distributed even without connecting the neutral wire between the IMAG and the primary star-point of the transformer.)
Metallic frames of the consumers (and the IMAG) must be earthed either through the neutral of the transformer secondary winding (TN-C-S earthing arrangement) or via a separate protective-earth conductor (TN-S arrangement).

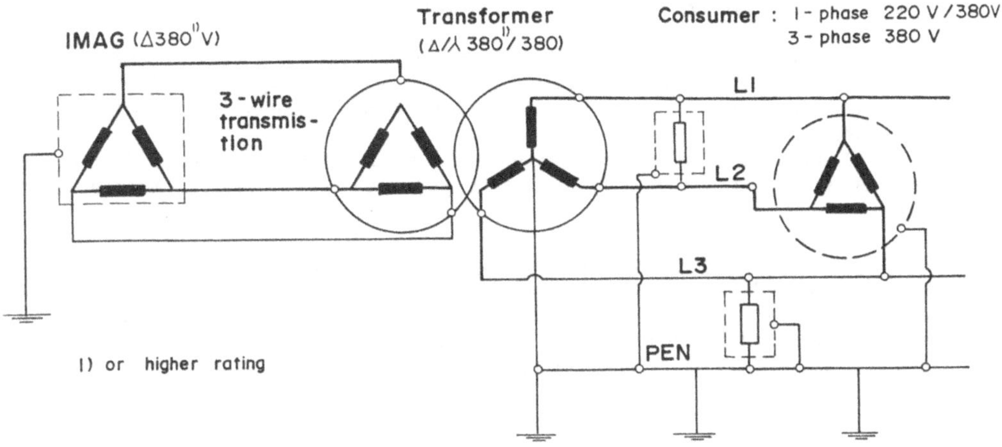

FIGURE 6.6 :
Internal connections and earthing arrangement of an isolated grid with IMAG and transformer (physical separation)

b) For small MHPs a transformer may be too costly; in order to eliminate to a large extent the effects of the third-harmonics, the **IMAG should be Δ-connected**. Such a system has however the disadvantage that only one voltage level is available for the customers.
Earthing can best be ensured by a common PE-conductor linking all frames of the appliances with earth at a number of points and with one phase of the IMAG (Fig. 6.7). With this system no excessive potential differences can occur between frames and non-electrical equipment of the customers.

The same lay-out is valid for a single-phase distribution using an IMAG with capacitors in C-2C connection.

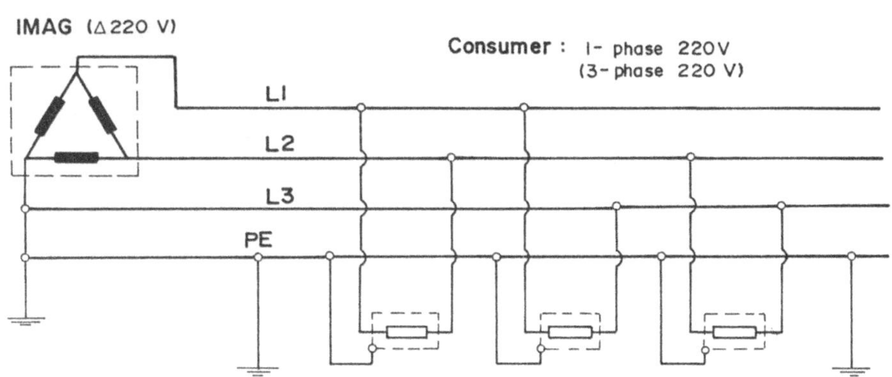

FIGURE 6.7 :
Internal connection and earthing arrangement of an IMAG without transformer

6.6.3 The IMAG - an inherently overload-safe generator

Overloading an induction generator (with fixed capacitors) does not create excessive current as with a synchronous machine. As shown in section 5.3, Figure 5.10, voltage and frequency will drop in an overload situation until the machine will finally de-excite without causing overcurrent. The IMAG is inherently overload safe.
The only situation where excessive currents can occur is when due to overspeed of the machine both voltage and current will exceed their limits. In this case the overvoltage protection will trip and shut down the system.

Overload protection of an **IMAG in isolated operation** can therefore be limited to normal fuses (or simple circuit breakers or relays) on all phases (including the capacitor circuit). Special arrangements to prevent nuisance tripping when motors are started direct-on-line (such as combined thermal and magnetic devices) are not necessary; the IMAG de-excites before an overcurrent condition can occur. In fact the overcurrent protection (fuses) in a stand-alone IMAG is simply a back-up to the overvoltage protection.

In **parallel operation** overcurrent can be produced by the grid. The IMAG should be protected against both short-circuit and overcurrents in the same way as a motor, e.g., with a combined magnetic and thermal relay in addition to the standard protection with fuses.

7. ECONOMIC CONSIDERATIONS

7.1 General

The main advantage of an IMAG is its low price as compared to any other generator, be it a purpose-wound induction machine or a synchronous generator. This is especially true for the small power range (below 10 kW) where synchronous generators are fairly costly and not readily available.
However, this cost advantage of the IMAG must be seen in the right perspective: in micro-hydropower plants in industrialized countries the investment costs for the electrical equipment are usually in the order of 5 to 10 % of the total investment (without costs of transformers and distribution network). A price advantage of the IMAG of say 50 % will therefore reduce total investment costs by maximum 5 % only. The spectacular price advantage of the machine has a rather limited effect on the overall investment costs.

In developing countries, where labour and building materials (stone, gravel, bricks, etc) are cheap, the distribution of costs of MHPs may differ considerably from that of industrialized countries. Most of the components of MHPs are nowadays locally manufactured except for the generator which is often the largest single component of the total costs because of import taxes, transport and unfavourable exchange rates. Installations where the generator accounted for 50 % of the total investment costs have been reported.
Using an IMAG (from local motor distributor or manufacturer/licensee) instead of an imported synchronous generator can make many MHP projects economically feasible and can trigger the support from financing agencies. When comparing costs of various generator options, the design engineer should include all investment costs (including turbine, civil engineering, etc.) and also the O&M costs in his analysis. Only in this way can the true economic benefit of an IMAG be evaluated.
O&M costs of the selected generator options might differ from each other because imported equipment will usually require imported spare parts which therefore incur additional costs throughout the service life of the plant.

7.2 Simple Methods of Comparing Project Costs and Benefits

In micro-hydropower installations of the lower output range (below 50 kW) the financial means of the implementing agencies are rarely sufficient to carry out a sound economic analysis. Instead of not treating economics at all, we propose hereafter a simple method of comparing project costs and benefits for both parallel and stand-alone electricity generation.

7.2.1 Identifying Project Costs

As mentioned above, the comparison between various generator options should be carried out on the basis of total costs and benefits including civil works, planning and design and O&M costs rather than on relative machine costs only. In a first step, all ***capital cost items*** of a project must be identified and grouped together according to their expected service lives (= period of time during which an investment facility can be used economically, usually equal or shorter than the technical lifetime). The service life of MHP components in developing countries is in the order of 10 to 20 years, the lower values for the generating units and the higher ones for civil works and services such as project planning and land acquisition. The following table provides an overview of the various capital cost items which might occur in an MHP.

TABLE 7.1 :
Capital cost items of an MHP with approximate values of service lives used for micro-hydro in developing countries

Cost Item	Service Life (years)
Mechanical Equipment (incl. transport, import taxes and installation costs): - turbine incl. mounting frame - transmission gearing (belts and pulleys or gear box) - (flywheel) - (speed governor)	10 to 20
Electrical Equipment (incl. transport, import taxes, installation costs): - IMAG or purpose-wound induction or synchronous generator - capacitance (for induction machines only) a) excitation of IMAG b) compensation of inductive loads in stand-alone plants - (load controller incl. ballast load) - switchgear (circuit breaker) - protection and monitoring equipment (fuses, relays, meters, etc)	10 to 20
Distribution networks a) low-voltage network (LV) b) medium to high voltage network incl. transformers (MV and HV)	10 to 20 20 to 25
Civil works - intake - head race and tail race channels/conduits - penstock - sedimentation chamber/sand trap - trash racks - control valves - powerhouse	15 to 25
Services - project planning - engineering design - topographical surveys - land acquisition	15 to 25

Each item of investment is converted into equal annual payments (annuities) using the so-called capital recovery factor. This factor permits calculating the equal instalments necessary to repay (amortize) a loan over the given period (the service life or credit duration) at the current market interest rate. The yearly payment is a varying combination of both interest payments and capital repayment. The recovery factor is defined as follows:

$$RF = \frac{i(i+1)^n}{(i+1)^n - 1} \qquad (7.1)$$

where i = interest rate (%/100) and n = period of loan or service life in years

If the project is not financed through loans (external financing) but by internal capital (savings, etc) of the owner/organization of the MHP, the same approach is used; i.e the current interest rate may be applied. This approach assumes that the owner could theoretically invest his capital on an alternative (the next best) use and thereby achieving at least the same interest.

Inflation causes a general rise in the price level and affects not only project costs and benefits but also the interest rate. Correcting the market interest rate with the current inflation rate will eliminate the fluctuating price level and make economic calculations much simpler since a constant price level over the whole project period can be applied. The corrected interest rate (to be used in Formula 7.1 above for the recovery factor) writes as follows:

$$i^* = \frac{(i+1)}{(a+1)} - 1 \qquad (7.2)$$

where i = market interest rate (%/100) and a = inflation rate (%/100).

In a next step, the costs of operation and maintenance (O&M) must be estimated. The following table may help to gather information of all possible cost items. As a general rule, total O&M costs without major replacements account for approximately 3 to 4 % of total investment costs of the MHP (using low wage levels without foreign personnel or mechanics/experts involved).

TABLE 7.2 :
O&M cost items of MHPs and approximate values of annual costs in % of capital costs

Cost Item	*Description*	*Yearly Costs in % of capital costs*
Personnel	<u>Operator</u>: Supervision of operation, preventive maintenance and inspection <u>Technicians, engineers</u>: General supervision, inspection and repairs in case of failures <u>Administration</u> (General manager, secretary) <u>Fitters & Labourers:</u> Inspection and preventive maintenance of distribution network, Replacement of faulty equipment <u>Labourers:</u> Repainting steel structures Desilting canals, forebay, storage pond	according to local conditions and wage levels
Office rents, office supplies and administrative costs	(without manpower)	according to local conditions
Electro-mechanical Equipment (PAT/Turbine, generator, governor)	Spare parts for the equipment, filters, lubricants, etc. (costs of material only)	0.6 - 2 % of electro-mechanical investment costs (covering costs of material only)
Civil Works (Intake, sand trap, canal, penstock, gates, trashrack, power-house)	Material for refilling of cavities, concrete for relining canals and intake works, paint for all steel structures, etc.	0.2 - 1 % of capital costs of civil works (covering costs of material only)
Distribution Network	Fuses, insulators, etc.	
Contingencies for major breakdowns	Flooding of power-house, intake etc.	in industrialized countries usually covered by insurance (yearly premium to be included)
Reserve for upgrading equipment	Replacing manual control with governor, etc.	depending on the technological level of the initial design
Depreciation allowance (only if investment costs fully or partly covered by subsidies and grants)		
Taxes and water rights, other duties		according to local rules and regulations

Total annual costs may now be expressed as follows:

$$Ak = K_o + (I_o - L) * RF + L * i^* \qquad (7.3)$$

Ak = total annual costs
K_o = O&M costs per year
I_o = investment costs
L = liquidation value of installations at the end of the recovery period (for MHP components where recovery period equals the service life: L = 0)
RF = recovery factor according to Formula 7.1 above (using the corrected interest rate i^*)
i^* = actual interest rate (market interest rate corrected with the inflation rate according to formula (7.2) above)

7.2.2 Identifying Project Benefits

A) Parallel Operation with a grid

It is not sufficient to compare annual costs of the different options (IMAG, purpose-wound induction machine or synchronous generators) to obtain the most favourable layout of the plant. Since IMAGs in general have a lower efficiency than the other options, the amount of electricity produced and delivered into the grid (national grid or mini-grid of various interconnected MHPs) is less and thus the annual income generated by the plant will be lower. Comparing yearly benefits rather than costs will therefore provide a more accurate criterion for the profitability of the technical solutions.

For parallel operation, the alternative to an IMAG is certainly a purpose-wound induction machine since these are less costly than synchronous generators and the excitation/magnetizing power required can (partly) be drawn from the grid without incurring additional costs. The difference between an IMAG and a purpose-wound induction generator mainly concerns the stator winding which is modified to avoid the saturation problems limiting the performance of IMAGs. Hence, the price difference between an IMAG and an induction generator is essentially determined by the costs for rewinding the stator. (In fact, the next step in the application of IMAGs would be the modification of the stator coils which is a rather delicate undertaking and cannot be treated within the scope of this manual.)

Yearly electricity production

Figure 7.3 shows the procedure of calculating the yearly energy production of an MHP operating in parallel to large grid (i.e. all energy can be absorbed). The individual steps are as follows:

a) - establish a flow duration curve from a hydrograph or from sporadic flow measurements
- determine nominal flow Q_n of the MHP; it has been found that the most economic layout point of an MHP operating in parallel is at a stream flow which is exceeded during 100 days per year (100-day rule)

b) - plot a head versus flow curve for the penstock pipes selected (hydraulic losses)

c) - plot the curve of turbine efficiencies versus flow; efficiencies at nominal flow can be taken from Appendix B8 while the efficiencies for part flow can be drawn approximately from Appendix B9

d) - calculate the mechanical power for several stream flows Q_i between Q_n and Q_{min}
- if a transmission gearing is used (V-belt, flat belt or gear box) use the approximate values of efficiencies η_{trans} indicated in Appendix B6 (for preliminary studies)

$$\boxed{P_{mec_g} = \rho\, g\, Q_i\, H_i\, \eta_{turbine_i}\, \eta_{trans}} \qquad (7.4)$$

where ρ = density of water (about 1000 kg/m^3)
g = gravitational acceleration (9.81 m/s^2)
Q = flow in m^3/s
H = net head in m

e) from Chapter 4 Selection, the efficiency of the IMAG at full and part load is known; for efficiencies of purpose-wound induction machine refer to manufacturers brochures

f) calculate the electrical power output of the IMAG (or other options):

$$\boxed{P_{el_g\,i} = P_{mec_g\,i}\, \eta_{g\,i}} \qquad (7.5)$$

g) calculate the yearly energy production which is represented by the area below the power versus time curve:

$$\boxed{E_{output} = \Sigma\left(\frac{P_{elg_i} + P_{elg_{i+1}}}{2}\,\Delta t\right)} \qquad (7.6)$$

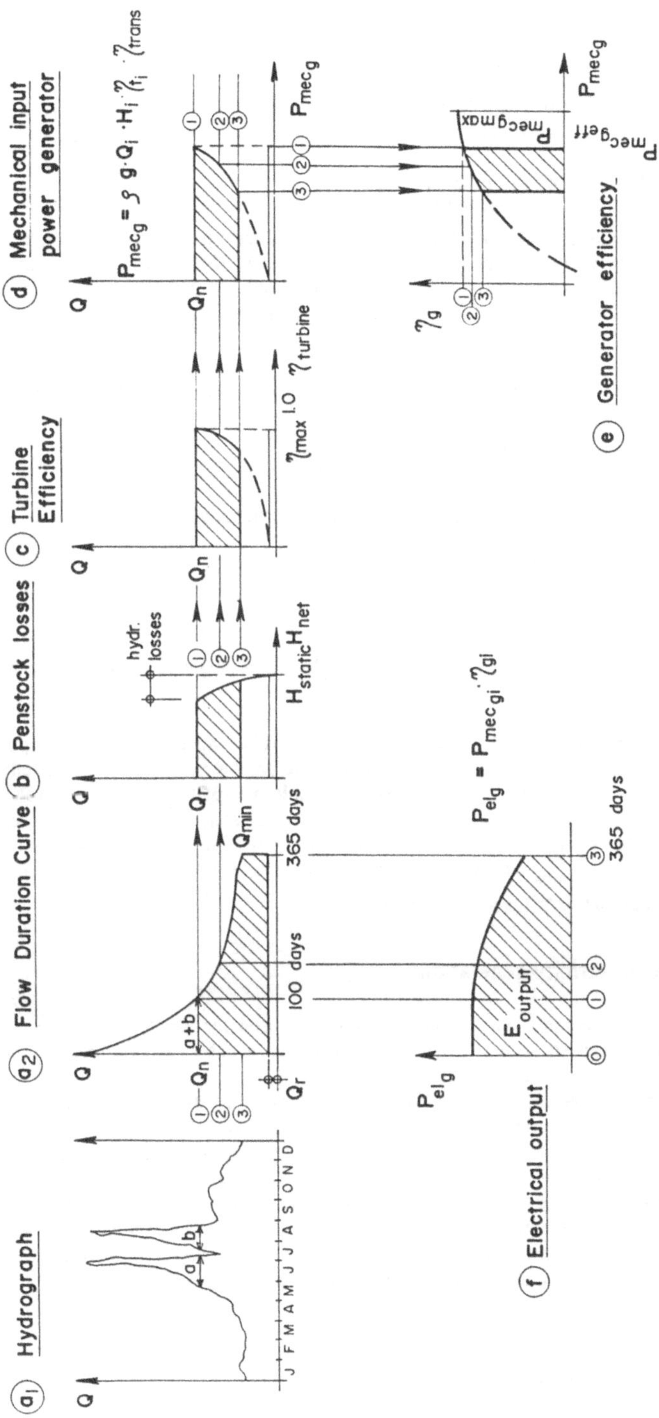

FIGURE 7.3 :
Calculation of yearly electricity output of an MHP in parallel operation

The economic viability of the different options can be evaluated in two ways:

a) calculating the unit production costs (currency / kWh):

$$\text{Unit production costs} = \frac{Ak}{E_{output}} \qquad (7.7)$$

where Ak = total annual costs (see Formula 7.3 above)
E_{output} = total energy production per year in kWh
The unit production costs or price per kWh represents the minimum price per kWh which the plant must receive from energy sales if cost and benefits are to break even at the end of the recovery period (credit duration or service life) or in other words if the installation should be economically viable. ***The technical option (IMAG or conventional generator) yielding the lowest unit production costs would be the economically favourable solution.***

b) Using the current price of energy (currency per kWh) and calculating annual income from energy sales:

$$\text{Annual income} = (E_{output} / a) * \text{current price} \qquad (7.8)$$

Comparing annual costs and income leads to the annual return generated by the plant:

$$\text{Annual return} = (\text{Annual income}) - Ak \qquad (7.9)$$

The option yielding the highest annual return should be chosen. A negative return means that the plant is economically not viable.

Overrating IMAGs as recommended in Chapter 4 Selection, might reduce total energy output and thus the annual return. It has been found that the loss on returns due to an overrated IMAG (lower efficiency at part load) is in the order of a few percent only (see worked example, Chapter 8). Selecting an IMAG without a safety margin i.e. which matches the turbine output power very closely, is therefore not necessary: even an overrated IMAG competes fairly well with other generator options. The proposed safety margin of 20 to 30 % (P_{mec_g} max = 1.2 - 1.3 P_{mec_g} effective) should be applied whenever possible as a safeguard against unexpected saturation problems of the selected IMAG.

B) Stand-alone electricity generation

Until now, synchronous generators have almost always been used for isolated operation. An IMAG will therefore have to be compared with a synchronous machine which will certainly be more expensive although it does not require excitation capacitors.
The trend in governing low-cost MHPs below 50 kW goes more and more towards electronic load controllers, available for both purpose-wound induction machines and IMAGs (IGC) and the synchronous machine (ELC). The generating unit will therefore operate continuously at nominal output and the drawback of the IMAG, the lower part-load efficiency, will therefore not become apparent. The only parameter which could influence the choice of the machine (apart from capital costs) is therefore the efficiency at nominal power. A higher efficiency means a greater electrical output at the available turbine flow and a greater number of households or appliances could therefore be served by the MHP. The analysis of project benefits for the different options of generating sets takes the following form:

a) calculate the nominal output of each generator option (use the generator efficiency η_g at the effective input power and not η_g max; see Figure 4.7 b)

$$P_{el_g} = \rho\, g\, Q_n\, H_n\, \eta_{turbine}\, \eta_{trans}\, \eta_g \qquad (7.10)$$

b) analyze the electrical loads to be connected to the generator; list the installed power of each appliance and their projected time of use during peak hours (usually in the evening between 7 and 9 pm)

Example: loads of a typical household in a newly electrified village

Load	Installed power P_{inst}	operating factor (% of time the appliance is in use during peak hours)	Installed power times operating factor
Lighting	80 W	90 %	72 W
Radio	20 W	60 %	12 W
Cooker (low- wattage)	200 W	60 %	120 W
Refrigerator (200 litre)	150 W	30 %	45 W
Total installed power	450 W	55 % <==	249 W

For an MHP serving a single household it is possible that all loads will be switched on simultaneously during peak hours; hence, the power output of the generator must be at least equal to the total installed power of the load. If the MHP serves a small village network with a larger number of households connected, it is unlikely that all appliances will be switched on simultaneously. Plotting a load curve (power versus time of the day, see Figure 7.4) according to the local habits and conditions and the projected use of appliances will provide an approximate value of peak power demand.

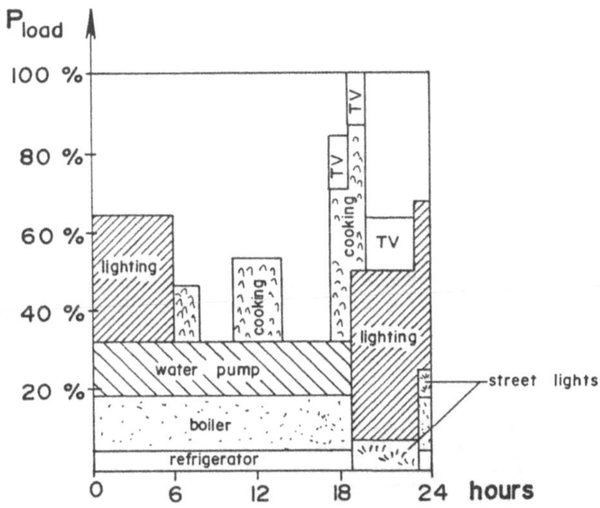

FIGURE 7.4 :
Example of a daily load curve of a small village electrification scheme

Plotting a load curve for a larger grid will become rather difficult. Based on statistical considerations the following formula expresses the simultaneity of electrical loads:

$$g_n = g_{inf} + \frac{1 - g_{inf}}{\sqrt{n}} \qquad (7.11)$$

where n = number of households connected
g_{inf} = factor of simultaneity for an infinite number of households

$$g_{inf} = \frac{\Sigma\,(P_{inst\,i} * \text{operating factor}_i)}{\Sigma\,P_{inst}} \qquad (7.12)$$

Peak power demand then becomes:

$$P_{load\,peak} = n * g_n * \Sigma\,P_{inst} \qquad (7.13)$$

For the example above, we obtain:

$$g_{inf} = \frac{249}{450} = 0.55 \text{ and with 40 households connected}$$

$$g_n = 0.55 + \frac{1 - 0.55}{\sqrt{40}} = 0.62$$

and finally peak load: $P_{load_{peak}} = 40 * 0.62 * 0.45 \text{ kW} = 11.2 \text{ kW}$

The relationship between the effective generator output and peak load is:

P_{el_g} effective > $P_{load_{peak}}$ + distribution losses + safety margin

c) It is obvious that with higher efficiencies of the generators, a larger number of households can be connected to an MHP thus increasing the income generated by the plant. The computation of the income depends on the tariff system applied. For small village grids with installed power of less than 1 kW, metering of actual consumption is too expensive. Many projects therefore apply a flat rate for a certain level of installed power. For a scheme with a single tariff level, the income increases proportionally with the number of households connected. The minimum yearly fee which the MHP has to charge to its customers to match at least the costs incurred is therefore given by:

annual fee$_{min}$ = Ak / n

where Ak = annual costs according to section 7.2.1 above
n = number of households connected (with equal values of maximum installed power)

d) The generator option (IMAG or synchronous generator) yielding the lowest annual fee would be the economically favourable solution.

Energy management

The example above shows that the capacity of the MHP is not well utilized since the nominal output is required only during the short period of peak demand while during the rest of the day, the available energy is largely dissipated either in a ballast load (load controller) or in the river-bed (flow controlled turbine). The relationship between actual and potential energy production/utilization is expressed with the station or plant factor StF:

$$\boxed{StF = \frac{\text{actual energy output of the plant}}{\text{potential energy output}}} \text{ during the same period of time}$$

Station factors for small hydropower installations are in the order of 10 to 30 % only. Improvements are possible by measures which flatten the load curve; this is usually referred to as energy management. In practical terms, this means that with a reduced peak load, a greater number of households can be connected which will in turn increase the income generated by the plant or increase its economic viability. Frequently applied measures for improving the station factor in small MHP schemes are the following:

- Base loads such as refrigerators or boilers are cut out during peak hours to make the necessary power available for other appliances such as cookers, lighting, fans, radios and TV sets.

- Applying lower rates for consumers operating during off-peak hours, thus making electricity more attractive

8. WORKED EXAMPLES

8.1 Parallel operation

Select an IMAG for the following site:
- H_{gross} = 30 m
- Q (flow duration curve see below)
- Penstock length = 66 m

8.1.1 Computation of net head

$H_{net} = H_{gross} - \Delta H_{losses}$

Penstock: steel pipe , internal diameter d = 200 mm , absolute roughness k = 1.0 mm

Friction losses (according to Moody) $\Delta H_{friction} = \frac{\lambda L}{d} \frac{v^2}{2g}$

Other losses = 10 % of $\Delta H_{friction}$

Discharge Q [l / s]	Net head Hnet [m]
90	25.5
70	27.2
50	28.6
40	29.1
30	29.5
25	29.6
22	29.7

Flow duration curve:

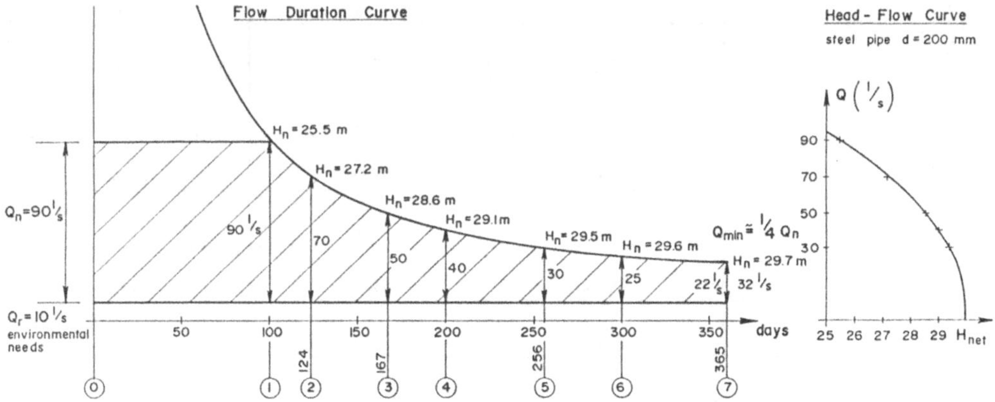

8.1.2 Turbine Selection

$P_{hydr} = \rho \, g \, Q_n \, H_n = 9.81 * 0.09 * 25.5 = 22.5$ kW

- from Appendix B7 : with $n_s = 1500$ rpm (direct coupled 4-pole IMAG, no transmission gearing)

  ==> Cross-flow turbine $n_q = 36$

- from Appendix B8: turbine efficiency $\eta_t = 70$ %

8.1.3 Mechanical input power of the IMAG

$P_{mec_g} eff = P_{hydr} \; \eta_t \; \eta_{trans} = 22.5 * 0.70 * 1.0 = 15.75$ kW

8.1.4 Approximate size of IMAG

- from Figure 4.6: $P_{mec_g} / P_N = 1.1$

  ==> $P_N = P_{mec_g} / 1.1 = 14.3$ kW

8.1.5 Review of manufacturers brochures

- 4-pole machines / $n_s = 1500$ rpm
- enclosure IP 44 or IP 54/55
- insulation class B
- nominal voltage 380 or 400 V
- mounting design: B3 (horizontal shaft, two end shields, on feet, free shaft extension)

	Leroy Somer	**ABB** (ASEA / BBC)	**Pemacco**
Designation/frame size	LS 160 L	160 L4 AG	160 L-4
P_N [kW]	15	15	15
I_N [A] at 380 V	29.3	30.5	29.8
I_{st} / I_N	5.8	6.5	7.3
T_{st} / T_N	2.4	2.5	2.3
T_{max} / T_N	2.5	3.4	3.0
T_{st} / T_{max}	0.96	0.74	0.77
n [rpm]	1445	1450	1460
η_m	89 %	88 %	89 %
$\cos\phi_m$	0.86	0.86	0.86

8.1.6 Selection parameters (according to Section 4.3.1)

- high motor efficiency η_m :	Pemacco and Leroy Somer: 89 %
- low slip / n close to n_s :	Pemacco: 1460 rpm
- high starting current:	Pemacco: 7.3 I_N
- low starting torque:	Pemacco: 2.3 T_N
- low ratio T_{st} / T_{max} :	Pemacco: 0.77

Conclusion: The Pemacco 160 L-4 seems to be the least saturated motor of this size

8.1.7 Maximum electrical output P_{el_g}

- from Figure 4.4 : (for $P_N = 15$ kW)

$$\frac{\sin\phi_g}{\sin\phi_m} = 1.30$$

with $\cos\phi_m = 0.86$ and $\sin\phi_m = 0.510$ ==> $\sin\phi_g = 1.3 * 0.510 = 0.663$ and $\cos\phi_g = 0.748$

Formula (4.3)

$$P_{el_g}\max = \frac{\cos\phi_g}{\eta_m \cos\phi_m} P_N = \frac{0.748}{0.89 * 0.86} 15.0 = 14.66 \text{ kW}$$

8.1.8 Maximum admissible input power

$$P_{mec_g}\max = P_{el_g}\max + P_N\left(\frac{1}{\eta_m} - 1\right) = 14.66 + 15.0\left(\frac{1}{0.89} - 1\right) = 16.52 \text{ kW}$$

$$> P_{mec_g}\text{eff} = 15.75 \text{ kW} \quad \text{okay}$$

8.1.9 Generator efficiency and power factor at part load

Part load values when generating similar to part-load efficiency and power factor when motoring:
- motor values to be plotted from Table 4.8; frame size 160

- generator values to be plotted similarly to motor curves but commencing with generator full load at

$$\frac{Pel_g \max}{P_N} 100\% = \frac{14.66}{15.0} 100 = 98 \% \text{ motor full load}$$

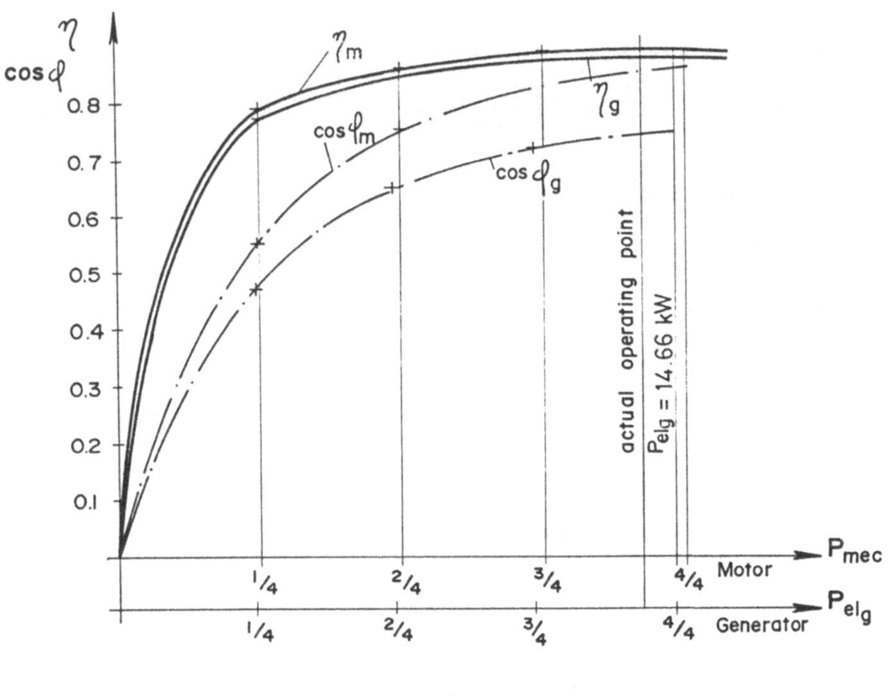

8.1.10 Efficiency versus mechanical input power

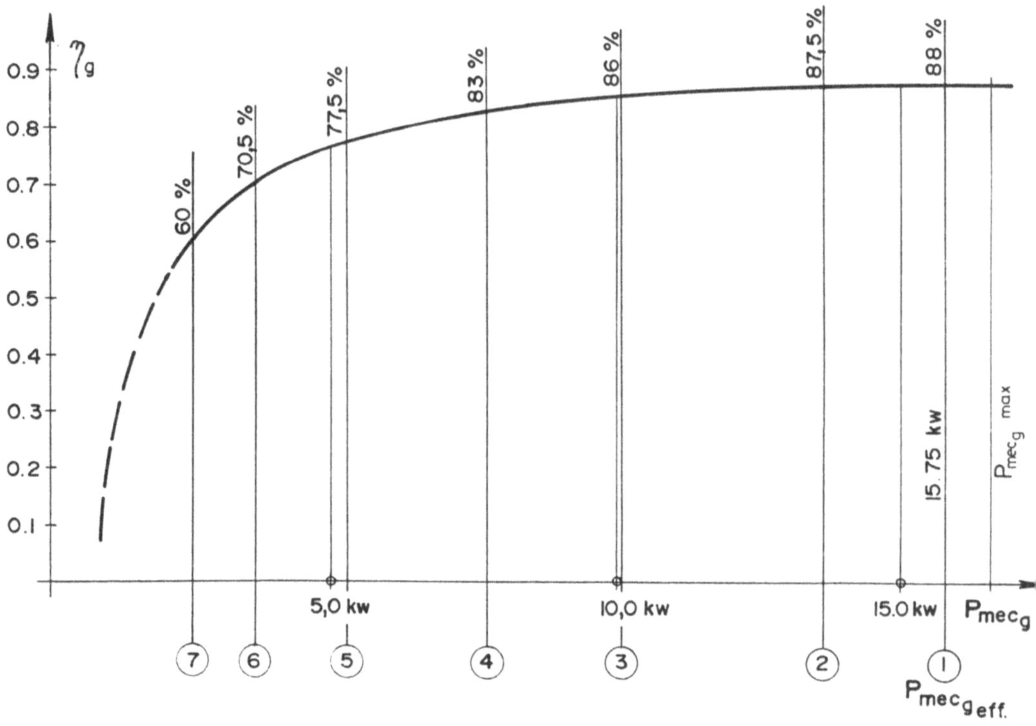

- from graph: efficiency at P_{mec_g} eff : $\eta_g = 88\ \%$

8.1.11 Effective electrical output at nominal flow

P_{el_g} eff $= \eta_g$ eff $* P_{mec_g}$ eff $= 0.88 * 15.75$ kW $=$ 13.86 kW

8.1.12 Capacitors required for parallel operation

The electricity utility requires reactive power compensation for $\cos\phi = 0.9$

- generator power factor for P_{el_g} eff = 13.86 kW : (from Figure 8.1.9) ==> $\cos\phi_g$ eff = 0.74

- Formula (5.2)

$\Delta Q = P_{el_g}$ eff $(\tan\phi_1 - \tan\phi_2) = 13.86\ \{\tan[\arccos(0.74)] - \tan[\arccos(0.9)]\} = 5.9$ kVar

- Formula (5.1)

$$C_{\Delta /\mathrm{phase}} \frac{\Delta Q}{3\ U^2\ 2\ \pi\ f} = \frac{5900\ \mathrm{Var}}{3 * 380^2\ 2\ \pi\ 50} = 43.4 * 10^{-6} = 43\ \mu\mathrm{F}$$

8.1.13 Alternative selection: purpose-wound induction machine

- Frame size : 180-4
- $\cos\phi_g = 0.84$
- $\eta_g = 89\ \%$
- $P_{el_g}\,\text{eff} = \eta_g * P_{mec\,g}\,\text{eff} = 0.89 * 15.75 = 14.0\ \text{kW}$
- Capacitors required for compensation to $\cos\phi = 0.9$

$$\Delta Q = P_{el_g}\,\text{eff}\,(\tan\phi_1 - \tan\phi_2) = 14.0\ \{\tan[\arccos(0.84)] - \tan[\arccos(0.9)]\} = 2.3\ \text{kVar}$$

$$C_{\Delta\,/\text{phase}} \frac{\Delta Q}{3\ U^2\ 2\ \pi\ f} = \frac{2300\ \text{Var}}{3 * 380^2\ 2\ \pi\ 50} = 16.6 * 10^{-6} = 17\ \mu\text{F}$$

Efficiency at part-load (purpose-wound induction machine)

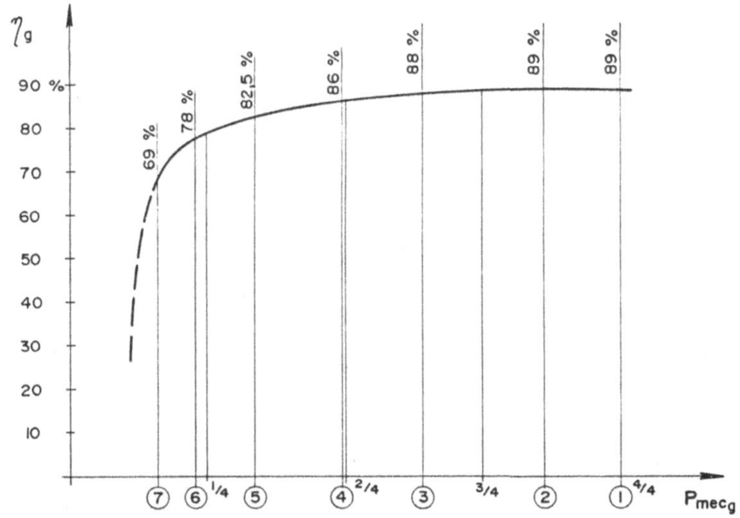

8.1.14 Yearly energy production

- Part-load operation of the Cross-Flow turbine (single-nozzle / non-segmented) (efficiencies from Appendix B8)

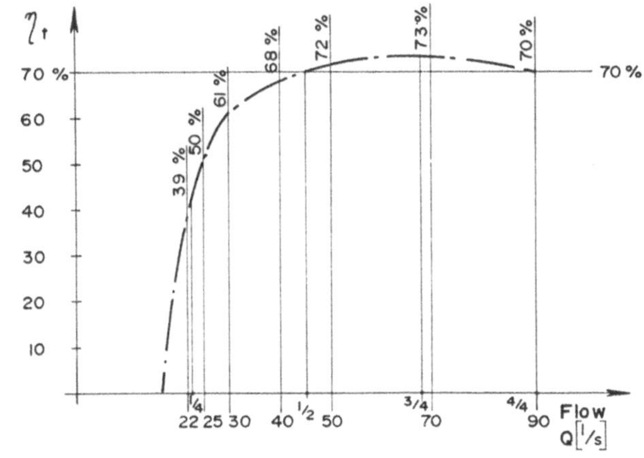

	Flow [l/s]	P_{mec_g} [kW] = $\rho\, g\, Q\, H_{net}\, \eta_t$
$Q_1 = Q_N$	90	15.76
Q_2	70	13.64
Q_3	50	10.10
Q_4	40	7.76
Q_5	30	5.30
Q_6	25	3.63
Q_7	22	2.50

8.1.15 Yearly energy production

$$E_{el} = \Sigma\, (P_{mec_g} * \eta_g * \Delta t)$$

P_{mec}	duration	IMAG (Pemacco 160 L-4)				Purpose-wound induction machine			
[kW]	[days]	η (%)	P_{el}	P_{el} (mean)	E [kWh]	η	P_{el}	P_{el} (mean)	E [kWh]
15.8		88	13.9			89	14.1		
	100			13.90	33'370			14.10	33'749
15.8		88	13.9			89	14.1		
	24			12.90	7'430			13.10	7'547
13.6		87.5	11.9			89	12.1		
	43			10.30	10'622			10.50	10'830
10.1		86	8.7			88	8.9		
	33			7.60	6'009			7.80	6'181
7.8		83	6.5			86	6.7		
	56			5.30	7'128			5.50	7'446
5.3		77.5	4.1			82.5	4.4		
	44			3.30	3'505			3.60	3'806
3.6		70.5	2.5			78	2.8		
	65			2.00	3'120			2.30	3'530
2.5		60	1.5			69	1.7		
	365				71'184				73'089

8.1.16 Economics

	IMAG (Permacco 160 L - 4)	Purpose-wound induction generator
	[US $]	[US $]
Purchase price	1600.-	6000.- (imported)
Capacitors a) 3 x 50 μF b) 3 x 20 μF	840.-	195.-
Wiring / switch for capacitors	300.-	160.-
El. equipment (protection, switch board, etc.)	500.-	500.-
Cross-flow turbine + water level control	8000.-	8000.-
Civil engineering, penstock	12000.-	12000.-
Design + contingencies	3000.-	3000.-
Total investment	**26'240.-**	**29'855.-**
Annuities: (service life: 15 years; i = 8 %; a = 5 %; i* = 2.86% recovery factor 0.083)	2178.-	2478.-
O&M costs per year	500.-	500.-
Total costs per year	**2678.-**	**2978.-**
Income: Production per year [kWh]	71'184	73'089
Price per kWh (to break even after 15 years) [$/kWh]	0.038	0.041
	100 %	**108.3 %**

8.2 Stand-alone operation

Select an IMAG for stand-alone electricity generation

$Q_{365} = Q_N = 90$ l/s $\quad H_{gross} = 30$ m $\quad H_{net} = 25.5$ m

load power factor $\cos\phi_{load} = 0.8$ at full load
Governing system : load controller (induction generator controller / IGC)

8.2.1 Selection

The IMAG suitable for this site is the same as in example 1 (Pemacco 160 L-4); additional capacitors are required to supply the reactive power for the load

P_{el_g} max = 13.9 kW (Pemacco 160 L-4)

- for excitation at max. output ($\cos\phi_g = 0.748$)

$Q_{c\,IMAG} = 13.9 \tan[\arccos(0.748)] = 12.33$ kVar

$$C_{\Delta\,/\text{phase IMAG}} = \frac{\Delta Q}{3\ U^2\ 2\ \pi\ f} = \frac{12330\ \text{Var}}{3 * 380^2\ 2\ \pi\ 50} = 90.6 * 10^{-6} = 91\ \mu F$$

- for compensation of inductive loads ($\cos\phi_{\text{load}} = 0.8$)

$Q_{c\ \text{load}} = 13.9 \tan[\arccos(0.8)] = 10.4$ kVar

$$C_{\Delta\,/\text{phase load}} = \frac{\Delta Q}{3\ U^2\ 2\ \pi\ f} = \frac{10400\ \text{Var}}{3 * 380^2\ 2\ \pi\ 50} = 76.4 * 10^{-6} = 76\ \mu F$$

these capacitors are in addition to those for the excitation of the IMAG at max. output

8.2.2 Alternative selection: Synchronous generator with electronic load controller (ELC)

- MARKON, three-phase B-range, industrial brushless with AVR
- Frame size 180 A/4 or B 354 A
- 50 Hz, $n_s = 1500$ rpm
- S = 20 kVA, $P_{el} = 16$ kW, $P_{mec} = 19$ kW

8.2.3 Loads of a typical home

type of load	installed power P_{el} [W]	operating factor (operation during peak hours) [%]	P_{el} * operating factor [W]
Lighting (flurescent)	80	90	72
Radio	20	60	12
Cooker	200	60	120
Refrigerator (compressor type)	150	30	45
Total installed power per home	450	55.3 <==	249

Statistically, the homes connected to the scheme will not use their total installed power all at the same time; peak load according to Formulae (7.11) to (7.13):

$g_\infty = 249 / 450 = 0.553$

$$g_n = g_\infty + \frac{1 - g_\infty}{\sqrt{n}} \quad \text{for } n = 40 \text{ homes:} \quad g_n = 0.553 + \frac{1 - 0.553}{\sqrt{40}} = 0.624$$

P_{load} max = n * g_n * P_{inst} / home = 40 * 0.624 * 0.45 = 11.2 kW

P_{losses} in distribution = 10 % of generator output = 1.4 kW

Generator peak load = 12.6 kW

< P_{el_g} max = 13.9 kW

8.2.4 Economics

	IMAG (Pemacco 160 L - 4) [US $]	Synchronous generator [US $]
Purchase price	1600.-	8500.- (imported)
Capacitors a) 3 x 100 μF (excitation) b) 3 x 80 μF (load)	1120.- 1100.-	
Wiring / switch for capacitors	350.-	
Load controller (IGC / ELC) incl. ballast load	2500.- (imported)	2500.- (imported)
El. equipment (protection, switch board, etc.)	500.-	500.-
LV distribution in the village (connecting 40 homes)	10'000.-	10'000.-
Cross-flow turbine (locally manufactured)	5000.-	5000.-
Civil engineering, penstock	12'000.-	12'000.-
Design + contingencies	3500.-	3500.-
Total investment	**37'670.-**	**42'000.-**
Annuities: (service life: 15 years; i = 8 %; a = 5 %; i* = 2.86% recovery factor 0.083)	3127.-	3486.-
O&M costs per year	750.-	850.- (incl. imported parts for syn. gen.)
Total costs per year	**3877.-**	**4336.-**
	100 %	**111.8 %**
Min yearly tariff per home (40 homes connected, P_{inst} max = 0.5 kW/home) flat rate (US $ / home)	**96.9** US$/home	**108.4** US$/home

APPENDIX A

BASIC THEORY OF ELECTRICAL ENGINEERING

1. INTRODUCTION

1.1 Electrical Circuit

Electric current consists of the flow of electrons in conductors. The energy of this flow can be converted into motion (motors) or heat (lamps, boilers). The three basic parts of an electrical circuit are the source, the conductor and the load. An electrical circuit resembles in many ways a hydraulic pipeline system.

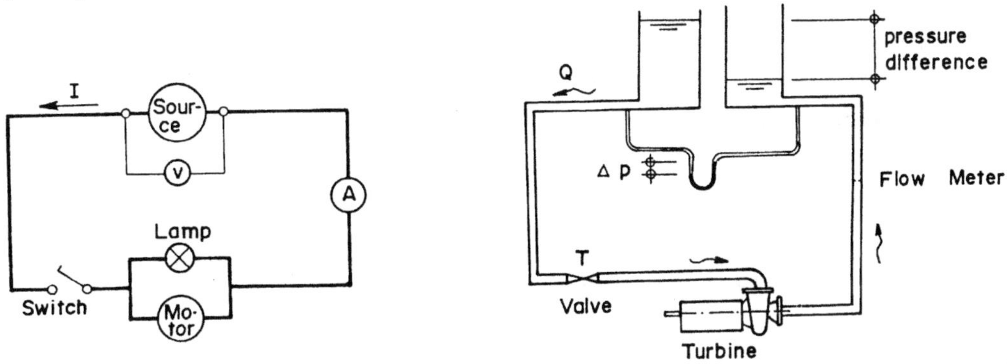

FIGURE A1 :
The electrical circuit and its hydraulic equivalent

1.2 Voltage

The electrons in the conductors move only as long as there is a source of electrical pressure available. Electrical pressure, known as voltage, is obtained from batteries, generators or photoelectric cells (which make up solar panels). Voltage is similar to the pressure difference (Δp) in the hydraulic system measured with a manometer.

Symbol for Voltage: U **Unit: Volt (V)**

Voltage is measured by means of a **VOLTMETER** which always measures the difference of electrical pressure **between two points** of an electrical circuit, i.e. the voltmeter is connected across (in parallel with) the source or the load.

The voltmeter is an instrument with a low consumption, or in engineering terms, with a high impedance (high resistance to the flow of electrons, so that the electrical pressure difference is not discharged through the voltmeter - see Figure A1 above).

1.3 Current

Applying voltage (pressure difference) to an electrical circuit, current flows in the direction of the pressure gradient. Current is the charge (charged electrons) flowing through a conductor per unit time (second) and comparable to the discharge (m3/s) in the hydraulic circuit.

Symbol for Current: I **Unit: Ampere (A)**

The current is measured by means of an **AMMETER** connected in series (in-line) between the source and the load; i.e. the current flows through it like in an orifice-plate flow meter in the hydraulic system of Figure A1.

The ammeter is an instrument with a low resistance to the flow of electrons (in engineering terms: low impedance). So the loss of pressure (voltage drop) across the ammeter is small. Note that the current in a circuit without parallel conductors or loads remains constant throughout the circuit.

1.4 Effects of Current Flow

Electric current flowing through loads produces the following effects which are used in various applications:

- **thermal (heat):** heating devices (boiler, tea kettle), tungsten lamps (the metal filament in a light bulb glowing white and giving heat and light, but does not burn due to the absence of oxygen), industrial furnaces, electric welding;

- **magnetic:** motors, relays, bells, electromagnets, measuring instruments (see also Section 3)

- **chemical:** electrolysis (separation of substances into its chemical components), accumulators, galvanization;

- **luminescent:** fluorescent tubes.

1.5 Power

Power is the rate of work done or energy supplied/consumed per unit time. One Watt is the power used when one Ampere is forced through a circuit by a pressure (voltage) of one Volt.

Symbol of power: P **Unit: Watt (W)**

Energy / unit time = Joule (J = W*s) / second = Watt

In a DC (direct current) circuit the power is always

$$\boxed{P = \text{Volt} * \text{Ampere} = U\,I} \qquad \text{(A1)}$$

The electrical power is measured by means of a **WATTMETER** which has two circuits (see Figure A2 below):

- voltage measurement (in parallel with the load or the source)

- current measurement (in series with the load and source)

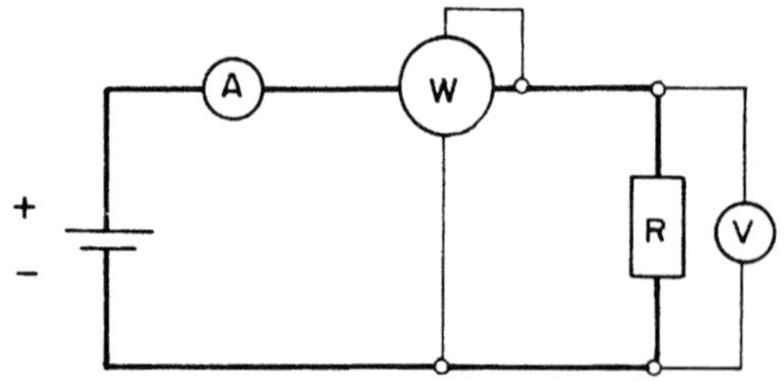

FIGURE A 2 :
Measuring Power with a Wattmeter

2. DIRECT CURRENT (DC)

If current does not change its direction of flow, it is called direct current and the symbol is:

= or -- (DC)

If a constant voltage is connected to the terminals of a resistor (lamp, heating device), direct current (I) will flow through that resistor.

Sources of direct current are:

- cells (batteries) and accumulators
- generators with commutator (commutator = mechanical device rectifying the direction of a current)
- diode bridge rectifier (diode = semi-conductor element which lets only one direction of current pass; thus an alternating current can be rectified (made direct current)
- thyristor converters (similar to diode but with regulation)

2.1 Resistance

2.1.1 General Formulae

Like friction losses in pipes of a hydraulic system, conductors oppose the flow of an electric current; this property is called resistance. All materials have some resistance. Those who offer high resistance to current are insulators.

Symbol of Resistance: R **Unit: Ohm (Ω)**

The resistance of a conductor to current flow depends on:

- the resistivity of the material ρ (Ω *mm2/m)
- the length l of the conductor (m)
- the cross-sectional area A of the conductor (mm2)

$$R = \frac{\rho \, l}{A} \qquad \text{(A2)}$$

The most common of all conductors used in electrical machines are copper and aluminium wire. However, resistance is not constant; it changes with temperature. The resistivity variation for copper and aluminium in function of temperature is linear.

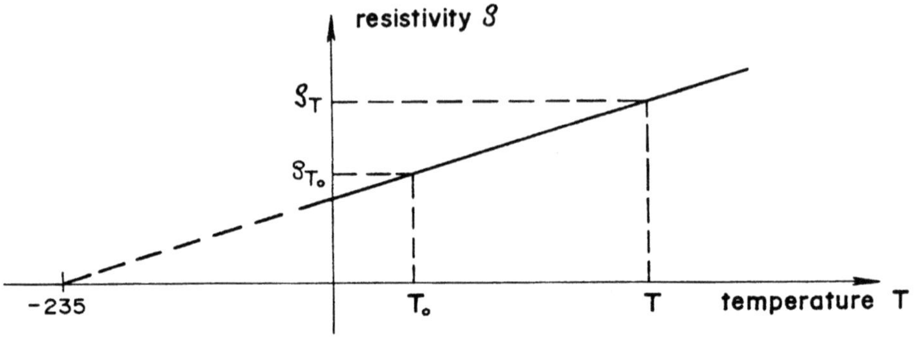

FIGURE A 3 :
Resistivity of copper as a function of temperature

For copper, the following formula applies:

$$\boxed{\rho_T = \rho_{To}\,(1 + \frac{T - T_o}{235 + T_o})} \qquad (A3)$$

(for aluminium, replace the value of 235 with 250)

ρ_T = final resistivity in Ω mm2/m

ρ_{To} = initial resistivity in Ω *mm2/m

T = final temperature in ^{o} Celsius

T_o = initial temperature in ^{o} Celsius

As initial conditions, the resistivity at $T_o = 20^0$ C is usually assumed; the resistivity of copper and aluminium for this temperature are:

- copper: $\rho_{20^\circ C} = 0.0175\ \Omega$ *mm2/m
- aluminium: $\rho_{20^\circ C} = 0.03\ \Omega$ *mm2/m

Since resistance is proportional to ρ, the above formula applies also for resistance at temperature T :

$$\boxed{R_T = R_{To}\,(1 + \frac{T - T_o}{235 + T_o})} \text{ for copper} \qquad (A4)$$

2.1.2 Ohm's Law

There is a definite relationship between the voltage U across a resistance and the current flowing through it, known as Ohm's Law:

$$\boxed{U = R\,I} \qquad (A5)$$

where U = Voltage across the resistance in Volt
I = Current in Amperes
R = Resistance in Ohms

2.1.3 Power Losses - Joule's Law

When a current flows through a resistance R, energy is lost in the form of heat (thermal effect). Power was defined as energy consumed per unit time and, in the case of DC circuits, is equivalent to voltage times current, $P = U * I$. These energy losses cause a decrease of the driving force (= voltage) of the electrical current in a circuit, known as <u>**voltage drop**</u>. Applying Ohm's law $U = R * I$, the amount of power lost through a resistance R can be calculated as follows:

$$\boxed{P = U I \quad \text{and} \quad U = R I} \qquad \text{(A6) thus}$$

$$\boxed{P = R I^2} \quad \text{Power P in Watt (W)} \qquad \text{(A7)}$$

Joule's Law

The corresponding energy is: $E = P t = U I \ t = R I^2 t$ (Ws)

2.1.4 Connection of Resistances

- Series Connection

So far, only one resistance has been considered. In most practical applications circuits consist of several devices, e.g. a lamp and its feeding conductors represent both resistors. In a **series circuit** the devices are connected one after the other along the conductor. Apparently, there is only one path for the current and the **same current flows through each resistor**.

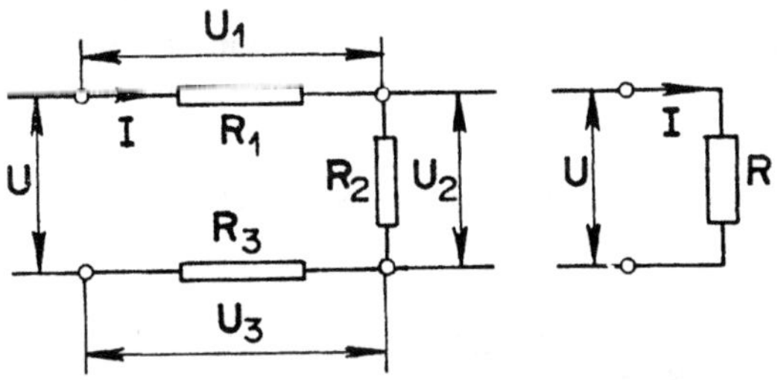

FIGURE A 4 :
Series Connection of Resistances

$$I = I_1 = I_2 = I_3 = \ldots = I_n$$

The total voltage U across the terminals of the source is equal to the sum of the voltage drops of each resistor:

$$U = U_1 + U_2 + U_3 + \ldots + U_n \qquad \text{applying Ohm's Law } U = R I$$
$$U = R_1 I + R_2 I + R_3 I + \ldots + R_n I = R I$$

Total circuit resistance or equivalent resistance R (in series connection)

$$\boxed{R = R_1 + R_2 + R_3 + \ldots + R_n} \qquad \text{(A8)}$$

- Parallel Connection

A parallel circuit offers a separate path for the current to each load.

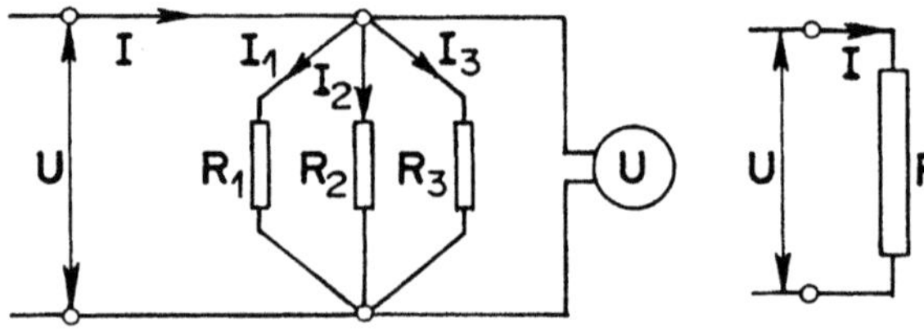

FIGURE A 5 :
Parallel Connection of Resistances

Each resistor is individually connected to the main source and therefore, each resistor operates under the same voltage as the source:

$$U = U_1 = U_2 = U_3 = \ldots = U_n$$

In a parallel circuit the resistors operate independently of one another and take current according to their resistances. However, the total current of all resistances is equal to the current I supplied by source:

$$I = I_1 + I_2 + I_3 + \ldots + I_n$$

Applying Ohm's Law I = U/R gives the total circuit or equivalent resistance R (for parallel connection):

$$\boxed{R = \frac{1}{1/R_1 + 1/R_2 + 1/R_3 + \ldots + 1/R_n}} \qquad \text{(A9)}$$

Example 1:

The wiring system of a house consists of the following devices:

- 3 lamps in parallel (R_1: 25 W at 24 V; R_2: 10 W at 24 V; R_3: 15 W at 24 V)
- 2 lamps in series ($R_4 = R_5$: 15 W at 12 V)

There is a 12 Volt DC supply at a distance of 200 m from the house. Copper wiring is used with a cross-sectional area of 1.5 mm2 (wiring inside the house between devices can be neglected). Calculate the total current in the supply line and the power requirement of the house if all devices are switched on simultaneously. Comment on the voltage across the lamps as compared to their ratings. Comment also on the size of the selected conductor (see Appendix B1 for standard wire sizes and maximum current). The ambient temperature is 35 ^{0} Celsius.

Solution
(see at the end of Appendix A)

2.2 Capacitance of Capacitors

2.2.1 General

Capacitors are a means to store electrical energy. As shown in section 1.2 above voltage (electrical pressure) is created by separating electric charge (electrons) into two different parts where one (the negative pole) apparently has a surplus of electrons and the other (the positive pole) has a lack of electrons.
If connected, the two parts will level out their different charges and a current (electrons) will flow from one pole to the other providing electric current to drive devices. All sources of electrical energy work on this principle.
Consider two metal plates placed at some distance from each other and supplied by DC voltage. The same separation of charge as in the source will be observed on the two plates: positive electric charge will accumulate on one plate and negative charge on the other. These two plates constitute an elementary capacitor.

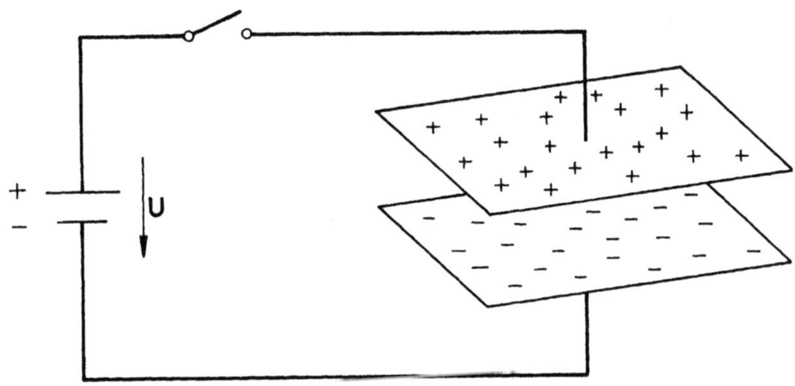

FIGURE A6 :
Elementary Capacitor

2.2.2 Capacitance

Symbol: C **Unit: Farad F = Ampere*second/Volt**

The quantity of electricity (charges) which a capacitor can store if supplied by a certain voltage depends only on the size (area of plates, distance in between) and the physical properties (filling of the capacitor) and is called **capacitance C** of the capacitor.

$$C = Q / U$$ where Q = quantity of electric charges in Ampere*seconds stored

In practice, capacitors have very low capacitance; it is therefore more convenient to use Microfarad (10^{-6} Farad) or Picofarad (10^{-9} Farad).

2.2.3 Variation of Current and Voltage

Charging and discharging of capacitors is not an instantaneous process, it takes some time until the charges (electrons) have moved to or from the capacitor. At the beginning of the charging process a current flow is observed while the current will stop as soon as the voltage across the capacitor has reached the same value as the source voltage.

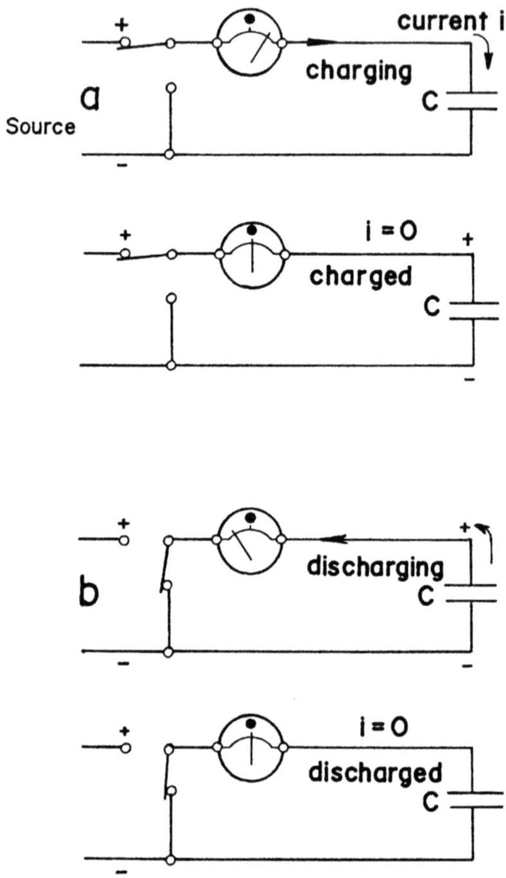

FIGURE A 7 :
Charging and Discharging a Capacitor

Applying a DC voltage to a circuit with a resistance R and a capacitance C (Fig. A 8a), it can be shown (using Ohm's law) that the initial current $i = U_0 / R$ will flow through R, where U_0 is the constant source voltage. Since this initial flow of current has created a voltage (charge) across the capacitor, the voltage drop (pressure to initiate flow) across the resistance is now lower than U_0, hence current i in the circuit will take a lower value. Thus neither i nor u take constant values during the charging process and we can write (note, small letters are used for current and voltage which vary in time):

$$i = C * \Delta u / \Delta t$$

The capacitor voltage increases in the same manner as the voltage drop across the resistance and therefore current in the circuit decreases; this is shown graphically in Figure A 8b.

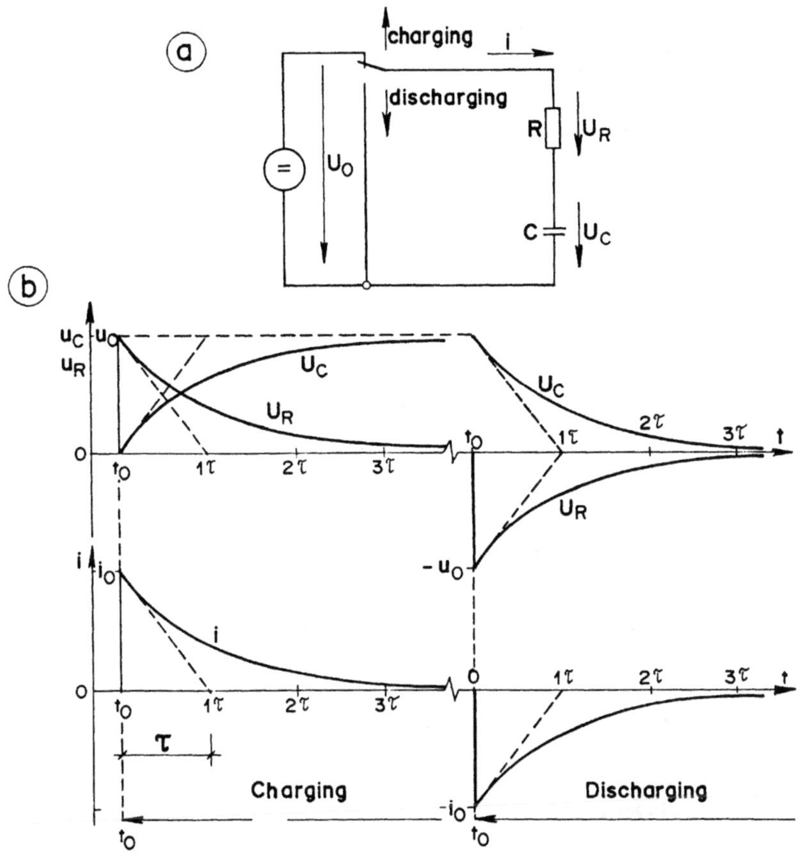

FIGURE A 8 :
Voltage and current variations when charging a capacitor

The charging and the discharging process follows an exponential law, i.e. circuit current and voltage across the resistor are maximum at the initial time t_0 and reach zero theoretically at infinity.
The duration of this charging and discharging process apparently depends on the capacitance and the resistor connected in series and is expressed by the **time constant τ :**

$$\tau = C * R \qquad (s)$$

(note the graphical interpretation of this value in Figure A 8b above)

2.2.4 Stored Energy of a Capacitor

The energy stored in a capacitor is the same as was supplied to it (source energy minus the energy corresponding to the ohmic losses). Referring to section 2.1.3 above we can write:

energy per period of time t: $e = u * i * t$ where u and i are the instantaneous voltage and current values according to Figure A 8 above.
Integrated over the whole duration of the charging process leads to

$$W = \frac{1}{2} C\, U^2 \quad \text{electrostatic energy stored in a capacitor} \qquad (A10)$$

in Ws (Watt*seconds)

Example 2: A photo-flash requires 100 Ws at a voltage of 800 V to produce a bright flash. Calculate the size of the capacitor required to store this energy. Determine also the resistor required to be connected in series with the capacitor if the current from a 1.5 V battery should not exceed 0.1 A when charging the capacitor.

Solution see at the end of Appendix A.

2.2.5 Connection of Capacitors

- Connection in series

Figure A 9 shows several capacitors connected **in series.**

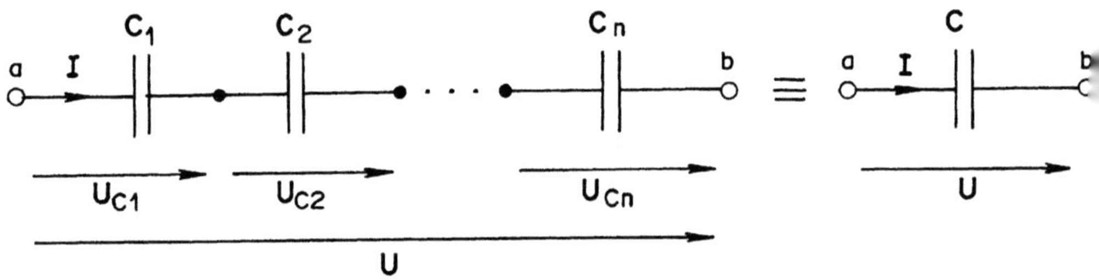

FIGURE A 9:
Capacitors in series

The total voltage U is equal to the sum of the individual voltage across each capacitor: $U = U_1 + U_2 + U_3 + ... + U_n$

Since there is only one path offered for the current to flow, I must be the same for all capacitors. From section 2.2.3 above we know that the current is directly proportional to the capacitance and the change of voltage through each capacitor: $I = C * \Delta U / \Delta t$. Hence, the product $C * \Delta U$ must be constant for all capacitors:

$I = C_1\, \Delta U_1/\Delta t = C_2\, \Delta U_2/\Delta t = C_3\, \Delta U_3/\Delta t = C_n\, \Delta U_n/\Delta t = C\, \Delta U/\Delta t$ where C is the equivalent capacitor of the serie.

since $I / C = \Delta U / \Delta t = 1/\Delta t(\Delta U_1 + \Delta U_2 + \Delta U_3 + ... + \Delta U_n)$ and $\Delta U = \Delta t\, I / C$ we can write:

$I / C = I / C_1 + I / C_2 + I / C_3 + ... + I / C_n$

Equivalent Capacitor C (for series connection)

$$C = \frac{1}{1/C_1 + 1/C_2 + 1/C_3 + ...\ 1/C_n} \qquad \text{(A11)}$$

If $C_1 = C_2 = C_3 = C_n$ the equivalent capacitor becomes $C = C_1/ n$

- Connection in parallel

Figure A 10 below shows several capacitors connected **in parallel.**

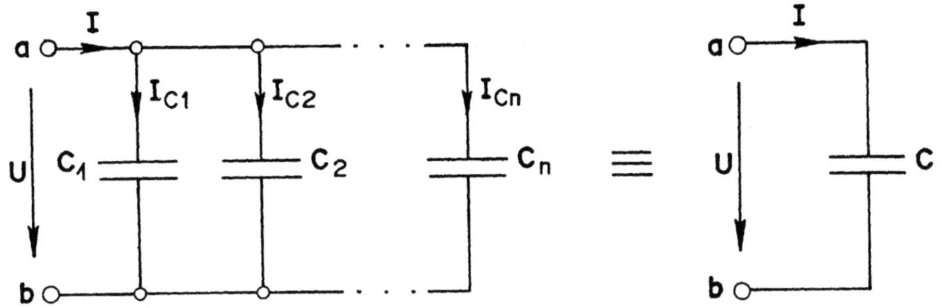

FIGURE A 10 :
Capacitors in parallel

Voltage across each individual capacitor is equal to the voltage of the DC source. The total current I is the sum of the individual currents flowing through each capacitor (note that current (electrons) actually cannot pass the capacitor; they accumulate on one plate of the capacitor whereas on the other a lack of electrons is created because on the latter, they have moved towards the source of the circuit.)
Applying the same relationship as for series connection above, the **equivalent capacitor for parallel connection** becomes:

$$\boxed{C = C_1 + C_2 + C_3 + ... + C_n} \quad \text{(A12)}$$

Example 3: The capacitor chosen for the photo-flash in Example 2 is to be replaced by two capacitors of which the second is twice as big as the first one.
Which connection is to be preferred if the capacitors should be as small as possible?

3. MAGNETISM - ELECTROMAGNETISM

3.1 Effects of Magnetism

If suspended, a permanent magnet always points with one end to the north. The reason for this is the natural magnetic field of the earth which is small but sufficient to attract for example the needle of a compass.
A permanent magnet produces around it a magnetic field where magnetic forces are active. These can be visualized by putting iron filings around the magnet bar. The filings immediately arrange into characteristic lines, the magnetic flux lines.

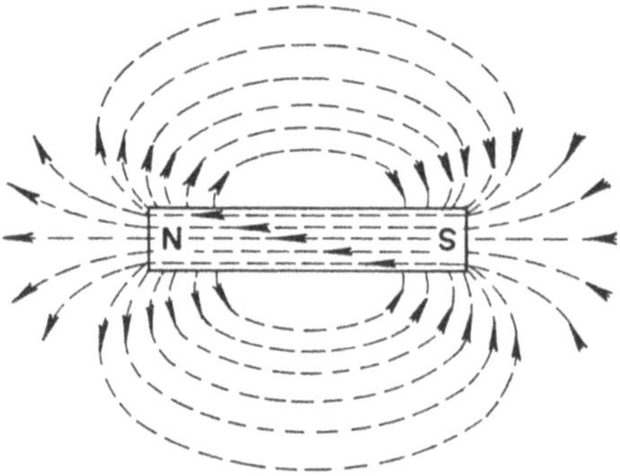

FIGURE A 11 :
Magnetic flux lines of a permanent magnet

It has been defined that magnetic flux lines flow from the **north to the south pole outside** of the magnet and from the south to the north pole inside it.

Similar poles repel each other while opposite poles attract each other.

A **piece of iron** in a magnetic field distorts the regular pattern of the flux lines. They concentrate in the piece of iron, because the iron offers them less resistance than air.

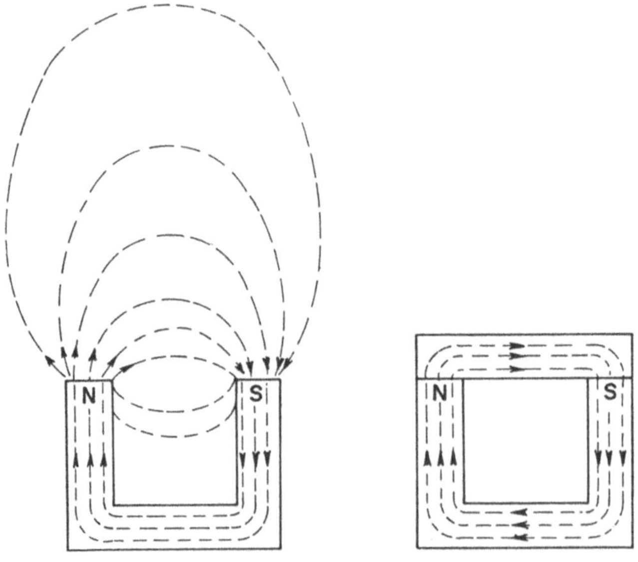

FIGURE A 12 :
Magnet flux lines in a closed/open permanent magnet

3.2 Magnetism and Electric Current

Magnetism is not a physical characteristic in its own right. Magnetism exhibits electrical effects and vice versa. When an electric current flows through a conductor, a magnetic field is created around it. This can be seen by placing iron filings near the conductor which carries current: the filings arrange around the conductor similar to the pattern around the permanent magnet (see Figure A 13a). The direction of the magnetic flux lines depends on the direction of the current. The following convention applies: A screw with a right-hand thread moves in the direction of the current if screwed in in the sense of rotation of the flux lines and vice versa.

a) single straight conductor **b) coil of n turns**

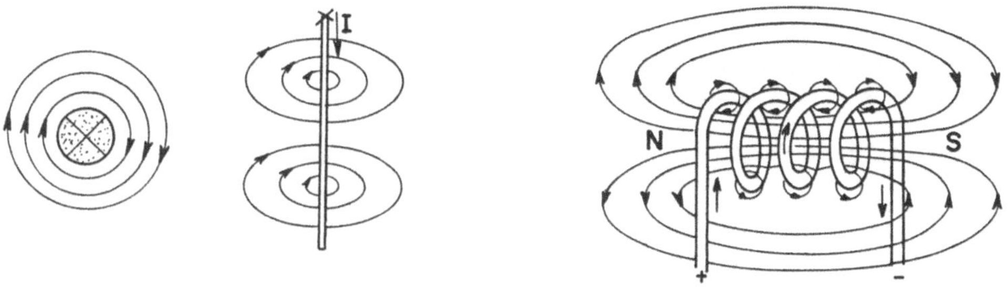

FIGURE A 13 :
Magnetic field produced by an electric current

The <u>magnetic field</u> around a straight conductor takes circular shape. The strength of the field is defined by the current flowing through the conductor and the area upon which it acts. The magnetic field strength diminishes with the square of the distance from the conductor.

Symbol of magnetic field strength: H **Unit: A/m**

Magnetic field strength H is a measure to describe the magnetic field created by the current in a conductor or coil (therefore unit amps).

A wire that has been coiled (Figure A 13 b), and with a current I flowing through it, also creates a magnetic field and each end acts like the poles of a bar magnet. Such a coil is called an electromagnet. The magnetic field strength H depends on:

- the current I flowing through that coil
- the number of turns N of the coil
- the mean length l of the flux lines

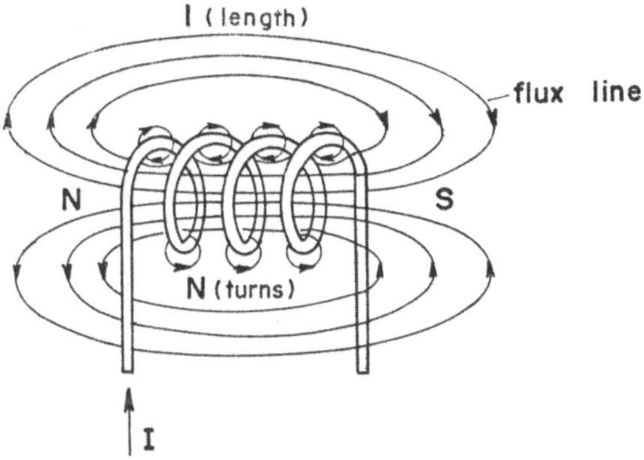

$H = N * I / l$ (A/m)

The strength of the field varies both in direction and magnitude: it is highest inside the coil and diminishes quickly outside of the coil due to the variable length of the flux lines. The constant product of the above formula N * I is therefore called **magnetomotive force or magnetic potential or total excitation θ**. Under its influence, work can be done such as arranging iron filings, turning the needle of a magnet etc.

3.3 Magnetic Flux and Magnetic Flux Density

When describing magnetic field strength we looked at the cause of the magnetic field or its potential due to the electric current. Magnetic flux on the other hand looks at the **potential effect of the magnetic field**. Therefore, the magnetic flux lines are considered to carry energy such as the stream-lines of flowing water.
A coil of N turns is wound around an iron core (or other ferromagnetic material, see Figure A 14). When an electric current flows through the coil, magnetic lines of force are produced. All these lines together constitute the magnetic flux.

Symbol of magnetic flux: Φ **Unit: Weber (Wb = V*s)**

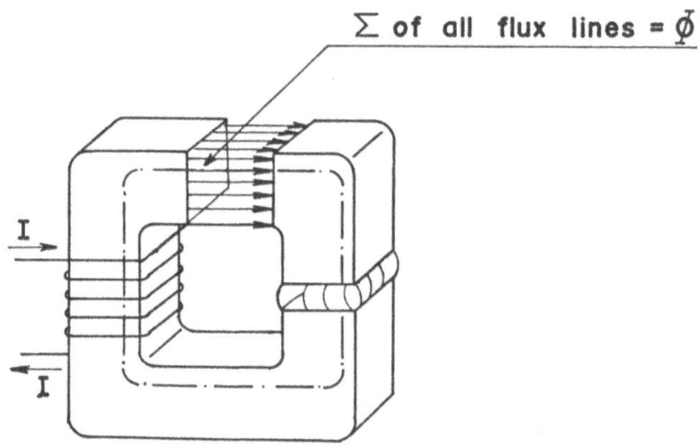

FIGURE A 14 :
Magnetic Flux

Magnetic flux per unit area is called magnetic flux density or magnetic induction.

Symbol of magnetic induction: B **Unit: Tesla (T = Wb/m2)**

$$B = \frac{\Phi}{A} \qquad (A13)$$

3.4 Magnetic Characteristic - Magnetization Curve

As shown in Figures A 12 and A 14 above, magnetic flux lines are concentrated in the material which offers little resistance such as in an iron core. For open cylindrical coils filled with air (Figure A13 b), the flux lines cover a much wider area and the magnetic effect of an electric current is therefore reduced. The magnetic field strength H is related to flux density B as follows:

$$B = \mu H \qquad \textbf{(A14)}$$

μ is the product of the **permeability in vacuum μ_0** and the **relative permeability of the material μ_r**:

$$\boxed{\mu = \mu_o \, \mu_r} \tag{A15}$$

$$\mu_0 = 4 * \pi * 10^{-7} \; (V*s/A*m)$$

$$\mu_r \text{ in air} = 1.0$$

The preferred material in magnetic circuits are ferromagnetic ones (= similar to iron). Ferromagnetic are iron, nickel and cobalt and some alloys of these. The permeability of ferromagnetic material is not constant. Figure A 15 shows the characteristic of a ferromagnetic material in comparison with air.

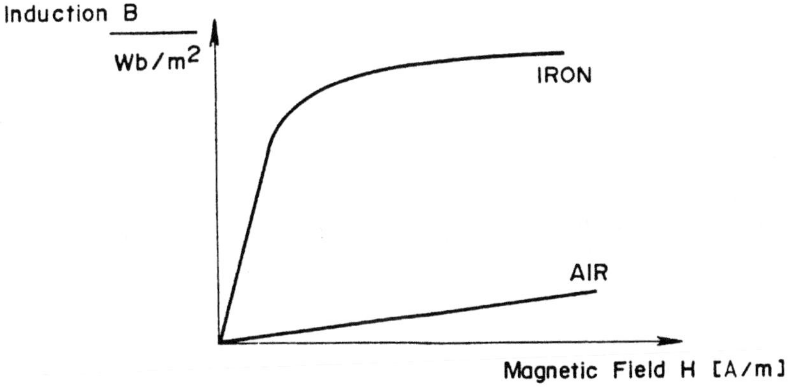

FIGURE A 15 :
B-H characteristic of iron and air

Above a certain point, the ferromagnetic quality seems to be lost and an effect called **saturation** occurs. Therefore, flux density or induction does no longer increase above a certain point even though the current in the coil (magnetizing force) is still increasing.

Figure A 16 below shows the complete characteristic of a ferromagnetic material. Complete means here that magnetization is carried through a complete cycle into negative values (change of direction of current in the coil). It reveals that the initial curve of magnetization (until saturation) is not retraced when the current in the coil (and subsequently the magnetizing force H) is decreasing; this effect is called **hysteresis**.

Note that the material, once magnetized, remains magnetic even after the electric current in the coil has stopped (H = 0). This is called **residual magnetism (or remanence B_r)** and is, for example, used to create permanent magnets.

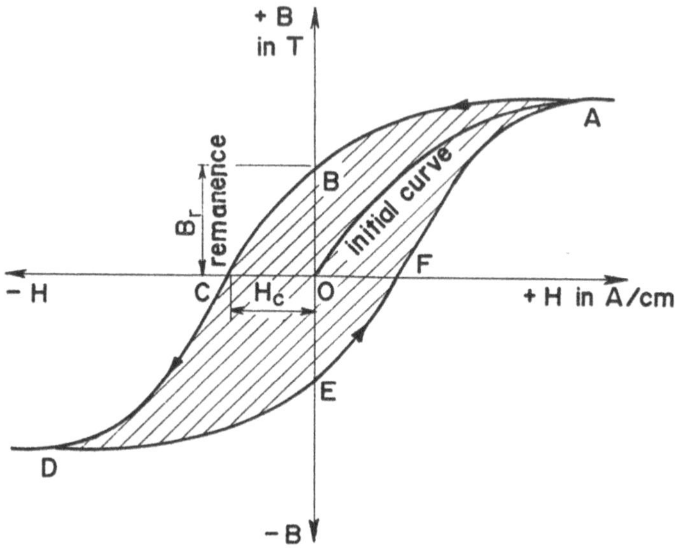

FIGURE A 16 :
Complete Characteristic of a ferromagnetic material

3.5 Generation of an Electromagnetic Force

Figure A 13 above has shown that a conductor carrying a current creates a magnetic field of circular shape around it. If another conductor is placed close to the first one, the following effects occur:

- two conductors carrying current in the same direction attract each other (see Figure A 17a);

- two conductors carrying current of opposite direction repel each other (see Figure A 17b).

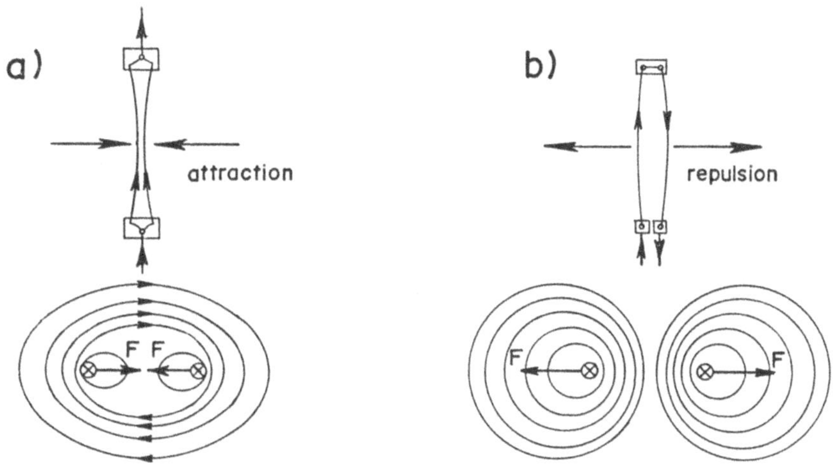

FIGURE A 17 :
Interaction of two magnetic fields

Superimposing the magnetic flux lines of the two conductors gives a resultant magnetic field with areas of high density and others of low density (see Figure above). A magnetic force acts on either of the two conductors towards the weaker area of the resultant magnetic field.

Similar effects can be observed when a conductor carrying a current I is placed in a magnetic field of density B (produced for example by a permanent magnet) and flux lines perpendicular to the conductor. A force is exerted on the conductor resulting in the displacement of the conductor.

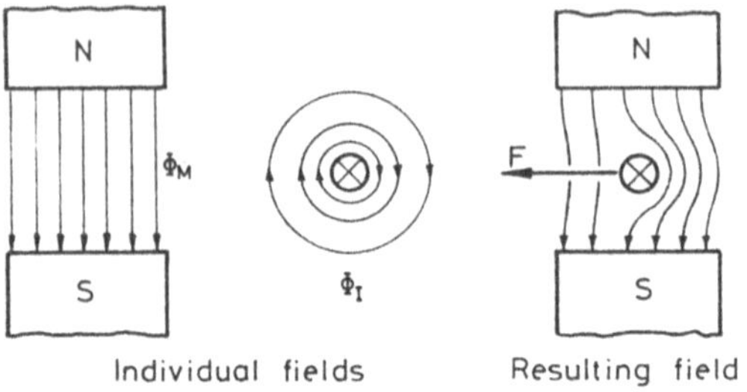

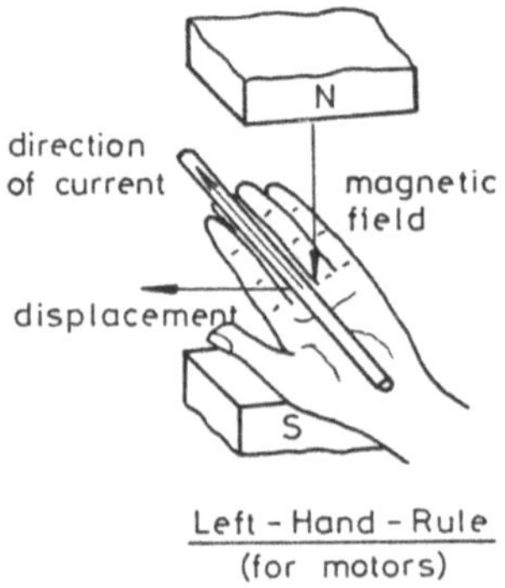

FIGURE A 18 :
Force on a conductor in a magnetic field

A qualitative solution of the problem is obtained by superimposing the two magnetic fields; the resultant field shows the areas of high and low density and the direction of the magnetic force on the conductor can be determined. Another qualitative solution is given in Figure A 18 using the "Left-Hand Rule".

The force is proportional to:
- the mean value of the magnetic induction B
- the current I flowing through the conductor
- the length l of the conductor in the magnetic field.

If the conductor is **perpendicular** to the magnetic field B:

$$\boxed{F = B * I * l} \qquad \text{(A16)}$$

As can be imagined, this phenomenon leads to the working principle of the DC motor. It will be treated in more details in the next section.

4. INDUCED VOLTAGE

4.1 General

Consider a permanent magnet producing a magnetic field of density B in the North to South pole direction. A conductor of length l is placed in the magnetic field perpendicular to the magnetic flux lines. The ends of the conductor are joined to form an electric circuit and a voltmeter is connected.
Moving the conductor at speed v through the magnetic field, the voltmeter shows a deflection. This voltage which causes a current to flow in the conductor is called the **induced voltage**.

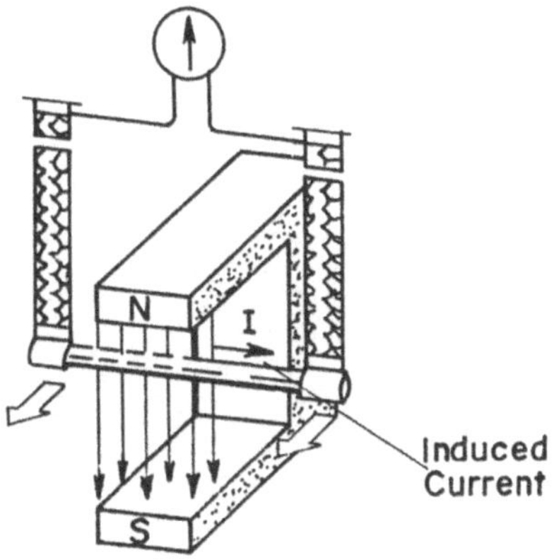

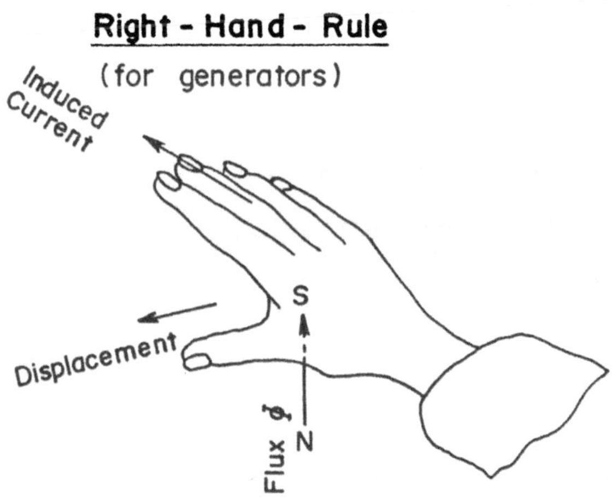

FIGURE A 19 :
Induced voltage

An interpretation of this phenomenon is as follows: electrons of the conductor move along with the conductor which is in fact like an electric current in a stationary conductor. This current creates a magnetic field around itself (see Figure A 13 above) which interferes with the magnetic field of the permanent magnet. The resultant field exhibits a magnetic force on the free electrons which therefore move to one end of the conductor while at the other end, a lack of electrons occurs. This separation of charges represents voltage, the induced voltage (see Figure A 20).

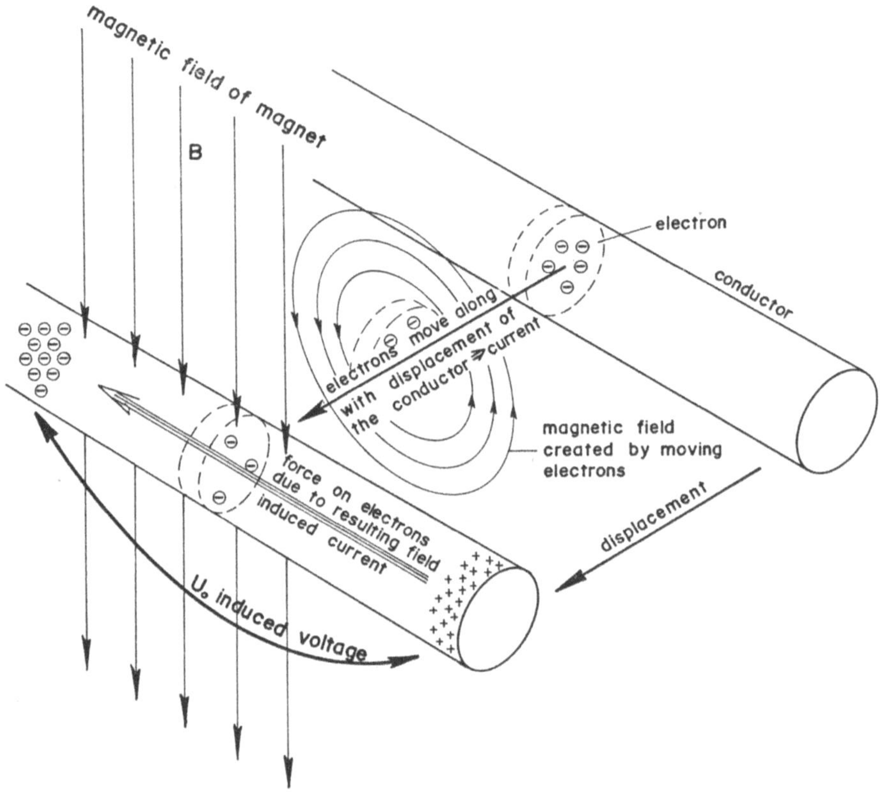

FIGURE A 20 :
Interpretation of Induced Voltage

The induced voltage is proportional to :
- the magnetic field density B
- the conductor speed v
- the conductor length l in the magnetic field

$$\boxed{U = B * l * v} \qquad (A17)$$

(Volt = Volt*s/m2 * m * m/s)

The direction of the induced current generated by the induced voltage (closed circuit) is defined by the "Right-Hand-Rule" (see Figure A19). Instead of a single conductor, a coil can be moved through the magnetic field which increases the induced voltage in the coil considerably.

In general, an induced current can be created either by:

- moving a conductor or coil through a magnetic field or

- changing the magnetic flux of the field with time (dΦ/dt)

The general law of induction writes as follows:

$$U_i = - N \frac{d\Phi}{dt} \tag{A18}$$

where N = number of turns of the coil
dΦ/dt = change of the magnetic flux with time

The negative sign indicates that the direction of the induced current due to an increasing magnetic flux flows opposite to the normal convention (right hand threaded screw).

The total voltage across a coil of resistance R is equal to the sum of the induced voltage U_i and the voltage drop due to the resistance R (Ohm's law):

$$u_{total} = R \cdot i + U_i = R \cdot i - N \frac{d\Phi}{dt} \tag{A19}$$

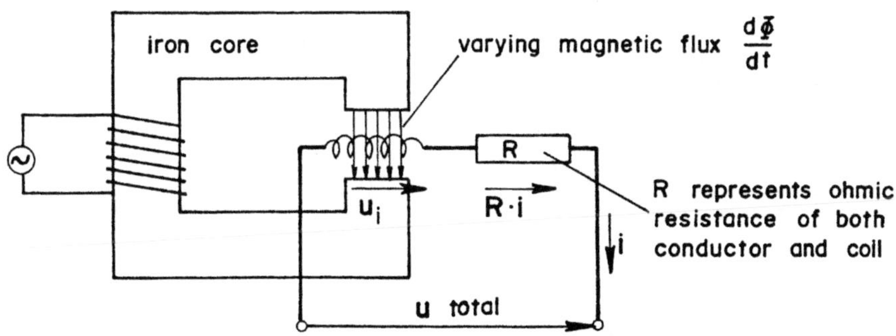

FIGURE A 21 :
Total voltage of a coil with induced voltage

4.2 Self-Induction

To show the phenomenon of self-induction, the following electrical circuit is considered: A DC source is placed in series with a switch, a lamp and a coil. A second lamp is connected in parallel to the source.

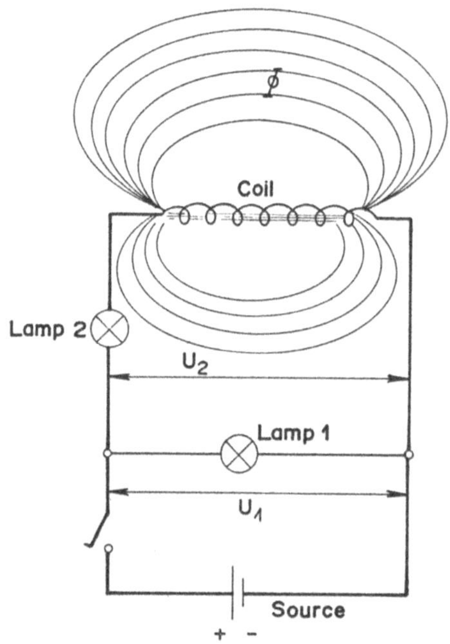

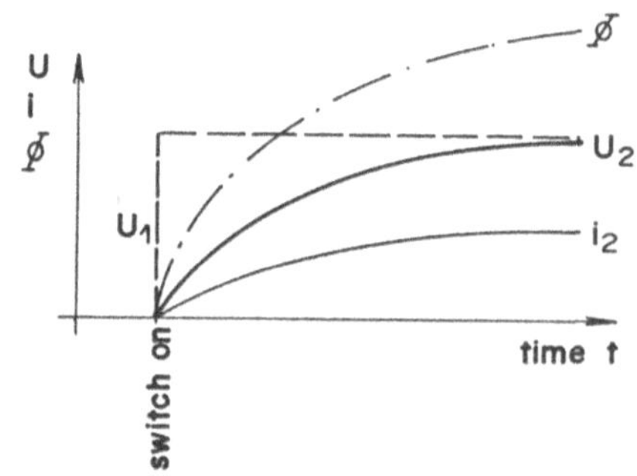

FIGURE A22 :
Self-induction

On closing the circuit, voltage and current are immediately available for lamp 1, while lamp 2 lights up only after a certain time. The reason for this lies in the magnetic field created by the current in the coil. Magnetic flux in the coil increases from zero to a maximum value corresponding to the current. This change of flux induces a voltage in the coil itself with a negative sign (see previous section) and thus opposes the voltage of the source. This prevents the immediate flow of current through lamp 2. For large electromagnets, this delay can last up to several minutes (coils/inductances will be treated in more detail in section 5).

4.3 Transformer Induced Voltage

Consider an iron core with two coils (see Figure A23 below). Coil 1 is connected to a DC source while the other is an open circuit containing a voltmeter. On closing the switch, a magnetic field is created through the iron core as a result of the current in coil 1. Due to self-induction, this current as well as the corresponding magnetic field reach only slowly (rapidly at the beginning then with gradually reduced acceleration) their end values.

This gradually changing magnetic field through the iron core induces voltage in coil 2 which has its maximum value right at the beginning and then reduces to zero at the time when the magnetic flux (Φ) of the core remains constant (remember it is only the **change** of magnetic flux which induces voltage). When opening circuit 1 with the switch S, the same change of magnetic flux occurs as when closing the circuit but the induced voltage in circuit 2 is now the reverse of the former value.

<u>**Conclusion:**</u> **Coils always show a certain degree of inertia to changing conditions. Current and voltage cannot rise or drop sharply in a circuit with a coil.**

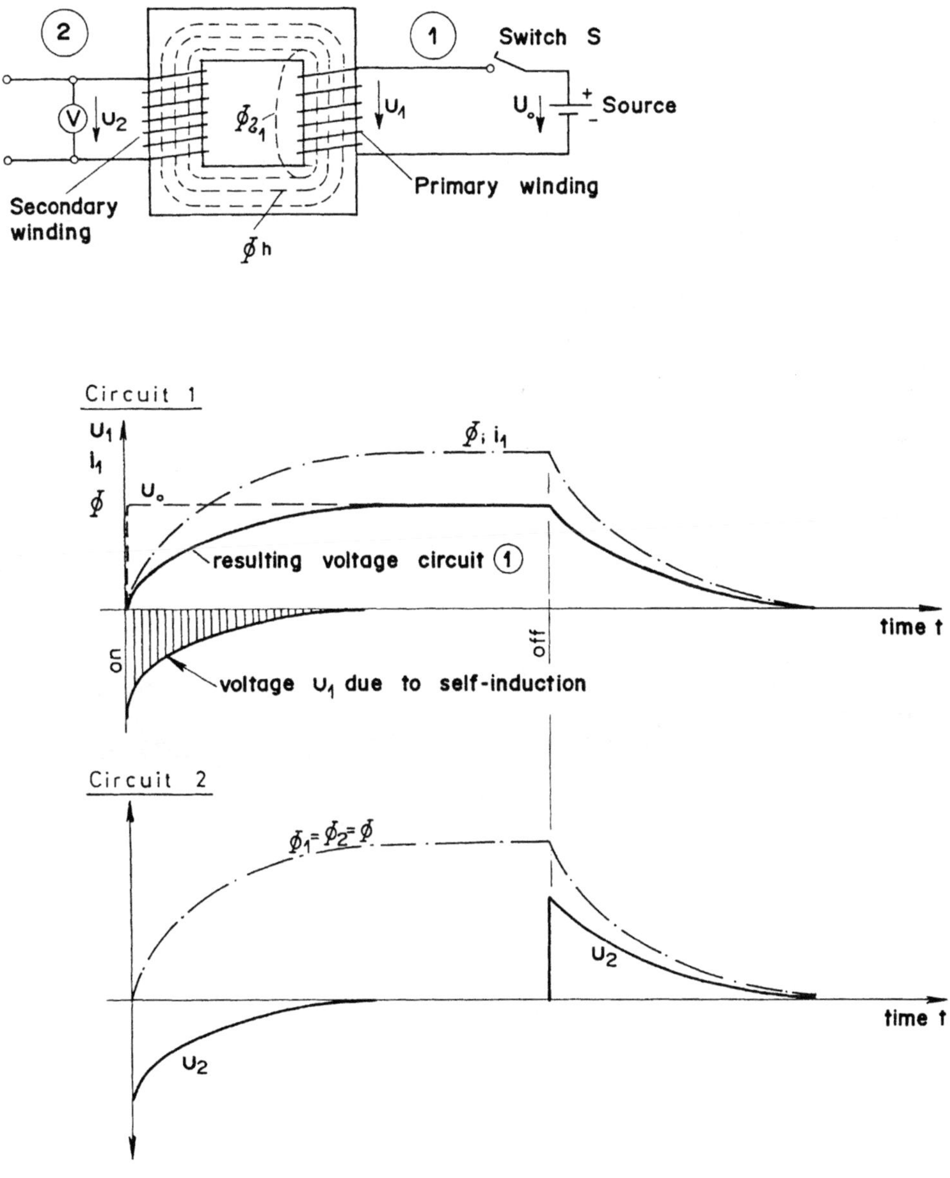

FIGURE A 23 :
Transformer induced voltage

If the switch S in circuit 1 (containing the primary winding) were opened and closed with a certain frequency, a sort of alternating current would be obtained in circuit 2 (containing the secondary winding), i.e. current and voltage would change direction continuously. A more convenient way to produce an interrupted current in circuit 1 is the use of alternating current (see also section 6).
As the name transformer implies, (alternating) current and voltage values can be transformed by means of a transformer into higher or lower values which are more suitable for a certain application. This can be achieved by chosing an appropriate ratio of coil sizes (number of turns) in the two circuits. For an ideal transformer we can therefore write:

$$\frac{u_1}{u_2} = \frac{N_1}{N_2} \quad \text{and} \quad \frac{i_1}{i_2} = \frac{N_2}{N_1} \qquad \text{(A20)}$$

4.4 Eddy Currents

A magnetic field of varying magnitude induces voltage in any material (except insulators) existing in the field. It is obvious that the iron core of a coil itself carries induced voltage and current. Due to the absence of a well defined circuit, these voltages are short-circuited within the core along the shortest possible path and produce high currents and therefore heat (losses). These currents are called eddy currents.
To prevent excessive eddy currents, the iron in magnetic circuits is composed of thin iron sheets laminated in between each other. The plane of lamination is most effective if parallel to the flux lines (since induced current intends to flow rectangular to the flux lines) thus confining the eddy currents to paths of small cross section and correspondingly high resistance.

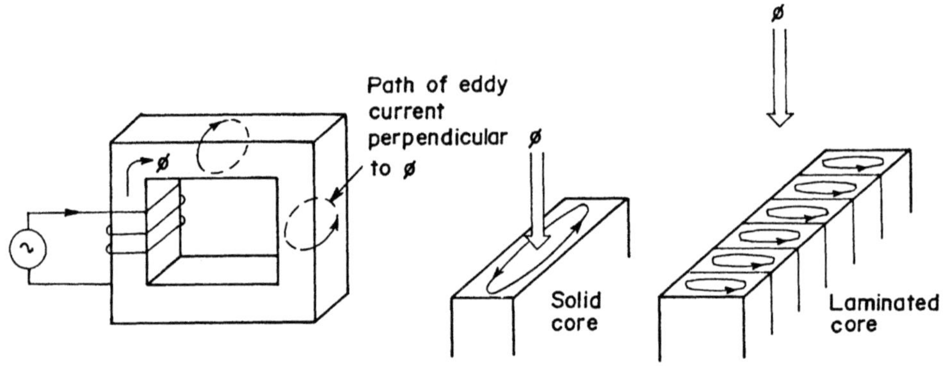

FIGURE A 24 :
Eddy currents and lamination of cores

5. INDUCTANCE

5.1 Definition

The inductance L of a coil is defined as the ratio between the magnetic flux Φ and the associated current I (see also Figure A23 above).

Symbol of inductance: L **Unit: Henry (H = V*s/A)**

$$\boxed{L = \frac{\Phi}{I}} \qquad (A21)$$

In Figure A 23 above it was assumed that the magnetic flux produced by the current of the primary circuit is equal to the flux in the secondary coil inducing voltage u_2. This is not the case in reality because there are flux lines which do not link the two coils but return through air and thus do not contribute to the induced voltage. The following definitions are therefore used:

Self-flux Φ_1 (total flux) produces self-induction with the ratio $\Phi_1 / I = L_1$ = Self-Inductance

Self-inductance is composed of (see Figure A25):

$L_{\sigma 1}$ = **Leakage Inductance** corresponding to the leakage flux $\Phi_{\sigma 1}$, and

L_m = **Mutual Inductance** which corresponds to the mutual flux Φ_m linking the two coils.

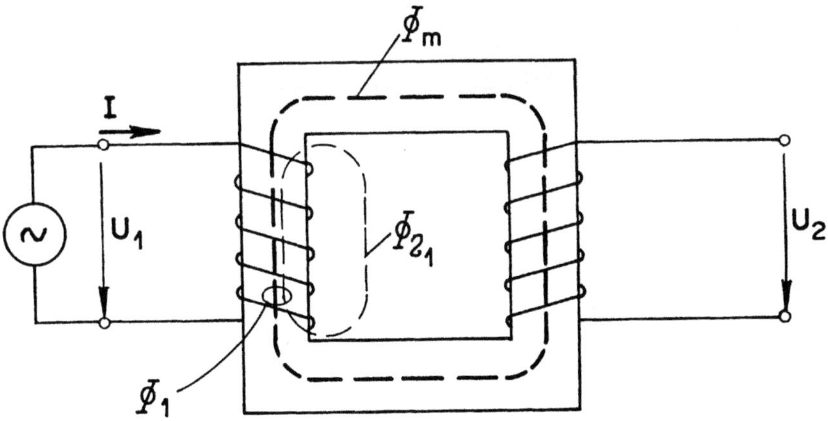

FIGURE A25 :
Self-flux, leakage and mutual flux

A coil has a resistance R and an induction L. These two elements can be assumed in series. The relationship between total voltage across a coil and current flowing through the coil is as follows (see also Figure A 21 above):

$u_{tot} = R\ i + L\ dI/dt$, using the definition of L we can write:

$\mathbf{u_{tot} = R\ i + N\ d\phi/dt = R\ i - u_i}$ where u_i is the induced voltage

This formula describes mathematically what was found by experiment above:
A coil having the inductance L tends to oppose a variation of current flowing through it which is, as we know, caused by self-induction.
When switching on a DC voltage source (U_0) on a circuit comprising a resistance R and an inductance L connected in series, the current grows according to an exponential law and reaches the end value $I = U_0 / R$. The time constant τ is defined as:

$\tau = L / R$ (s) compare with time constant for capacitors

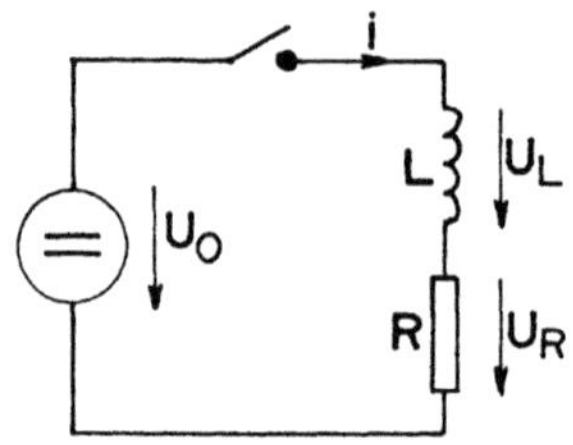

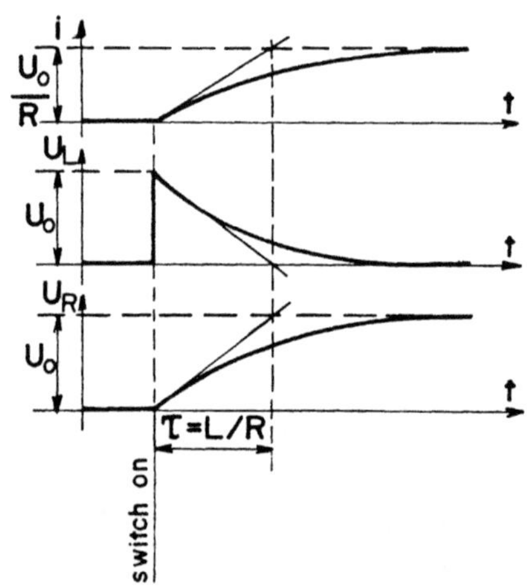

FIGURE A 26 :
Development of voltage and current across an inductance

Voltage u_L across the inductance has the initial value U_0 which is, according to the above formula, equal to the self-induced voltage u_i (because current i is stili zero and therefore the ohmic part R * i as well).
Voltage across the resistance is initially zero because source voltage and self-induced voltage of the inductance equal each other out. Final voltage across the resistor is U_0. For DC current it is then similar to a circuit with the resistor R only because L is zero due to no current variation (DC).

The inductance L stores a magnetic energy of:

$$\boxed{W = \frac{1}{2} L I^2} \qquad \text{(Joule J)} \qquad \text{(A22)}$$

This energy is stored in the magnetic field of the coil. It is released in form of electric current after opening the electric circuit. (Compare with the electrostatic energy stored in the electric field of a capacitor.)

5.2 Connection of Inductances

5.2.1 Inductors in series

Connecting inductors in series leads to the same formula as with resistances. Current I is constant through all inductors since the circuit offers only one path for the current to flow.

$$\boxed{\text{Equivalent Inductor } L = L_1 + L_2 + L_3 + \ldots + L_n} \quad \text{for series connection} \qquad \text{(A23)}$$

5.2.2 Inductors in parallel

As with resistances, the voltage across inductances connected in parallel is the same. This leads to the equivalent inductor for a series of inductors connected in parallel:

$$\boxed{\text{Equivalent inductor } L = \frac{1}{1/L_1 + 1/L_2 + 1/L_3 + \ldots + 1/L_n}} \quad \text{for parallel connection} \qquad \text{(A24)}$$

6. DC GENERATORS AND MOTORS

6.1 Working Principle of a DC Generator

So far induced voltage was obtained by varying the magnetic flux crossing a stationary conductor or coil. The same effect is created when a coil of N turns **rotates** in a constant magnetic field of a permanent magnet or an electromagnet. An elementary two-pole DC generator is shown in Figure A 27:

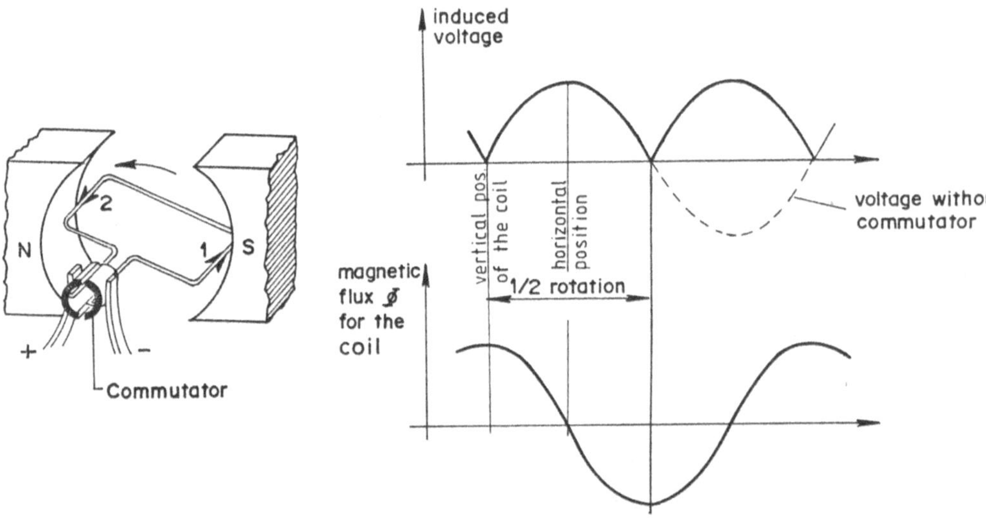

FIGURE A 27 :
Elementary DC Generator

The armature or rotor winding (here only a single rectangular coil) rotates in the magnetic field of a permanent magnet with north and south poles. Similar to Figure A 19 above, voltage is induced in the two conductors. If a load is connected to the winding, a current will flow in the circuit. From the rotating winding, this current is supplied to the commutator mounted onto the shaft of the machine. The commutator consists of two segments or collectors. Two brushes (made of carbon) ride on the commutator and transmit the current from the rotating shaft to the terminals of the machine.

The commutator serves to rectify the induced voltage and current in the rotor winding. Without this device the induced current and voltage would change direction if the coil of Figure A27 is in the vertical position.

As indicated previously, induced voltage is maximum if the conductors of the single coil move perpendicular to the magnetic flux lines; for the elementary DC generator, this will be at the horizontal position of the coil in Fig. A27. Similarly, induced voltage is zero when the coil is in the vertical position; i.e. the conductors do not cut flux lines. Therefore, voltage (and the corresponding current following it) are not constant; they vary periodically (in sinusoidal waves as we will see later) between a maximum and zero. To make better use of the magnetic field, an armature winding consisting of many coils instead of a single turn are used and the pulsating amplitude of voltage and current are reduced.

The use of permanent magnets to create the magnetic field across the rotor, known as excitation, is only possible for small generators. More common is the use of an electromagnet consisting of several field or stator coils on an iron core known as field poles. The stator winding can either be supplied by an external DC source (generator or a battery) or, as is usually the case, by the generator itself which is then called a self-excited generator. To start the induction process, self-excited DC generators can make use of the residual magnetism available in the stator poles as explained in section 3.4 above.

6.2 Working Principle of a DC Motor

A DC motor is basically a DC generator operated in reverse. The rotor windings are supplied by current through the brushes riding on the commutator. While flowing through the rotor windings, the current produces a magnetic field which interferes with the magnetic field of the stator. Referring to section 3.5 above, we know that the resulting field creates a force couple on the rotor which provides the torque for the rotation (see Figure A 28). Note that either the stator field or the current supplied to the rotor must change direction in order that the motor can rotate a complete turn.

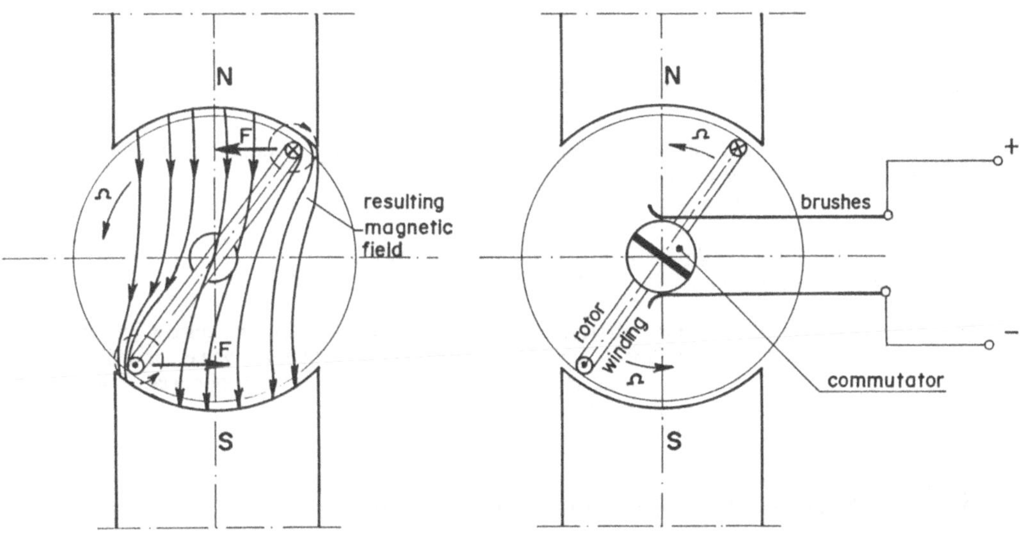

FIGURE A 28 :
Elementary DC Motor

Nowadays the DC motor is only utilized for applications with the following demands:

- battery-driven devices (movie cameras, toys, machine tools)
- adjustable motor speed over a wide range

7. ALTERNATING CURRENT (AC)

7.1 Advantages of Alternating Current

Alternating current is used in most industrial or household distribution systems for the following reasons:

- lower costs for high voltage transmission as compared to direct current;

- transformation of alternating voltage and current from one level to another is easy (transformer)

- AC motors and generators are of simpler construction than DC ones (no commutator); moreover, they are more economic and require less maintenance effort;

- cutting off AC is easier than cutting off DC (when disconnecting wires (switch), sparkover is automatically stopped in AC due to the alternating waveform (zero passage).

7.2 Characteristics of the AC signal

A signal is alternating when it varies in function of time around a mean value. Its direction changes continuously from positive to negative.
In electrical energy distribution systems:

- the variation is **SINUSOIDAL** and the mean value is zero;
- the variation is periodic, i.e. it is repeated within equal time periods.

The previous section showed that all generators basically produce alternating current in sinusoidal waveform. Figure A29 below gives the explanation. According to section 4.1, voltage is induced in a conductor moving perpendicular to a magnetic field. The induced voltage is proportional to speed and length of the conductor and the strength of the magnetic field B.

$$u = B * l * v_{\perp}$$

where $v_{\perp}$ is the speed of the conductor perpendicular to the flux lines.

However, the rotor winding of the generator in Figure A 29 moves only in its vertical position exactly perpendicular to the flux lines. In all other positions the horizontal velocity (the component contributing to the induced current) is lower and reaches zero when the coil is in the horizontal position. The velocity vector diagram shows the relationship between rotational speed and horizontal velocity of the coil which is obviously a sinusoidal relationship. The above formula therefore writes:

$$\boxed{u = B\ l\ v\ \sin\alpha} \qquad \text{(A25)}$$

where u = instantaneous voltage
v = peripheral velocity of the rotor coil

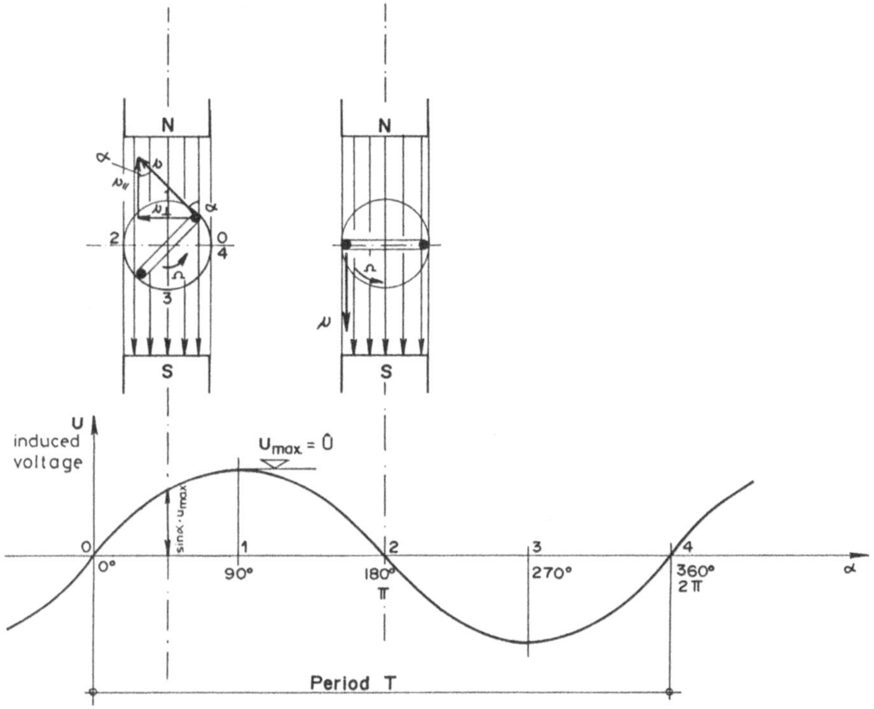

FIGURE A 29 :
Sinusoidal AC and its development

- **PERIOD**

Symbol: T **Unit: Second (s)**

The period is the time to complete one entire cycle, i.e. a negative and a positive half-wave. The rotor of the elementary generator in Figure A29 completes a full period after a **rotation of 360^0 or 2*π radians.**

- **FREQUENCY**

Symbol: f **Unit: Hertz (Hz)**

The frequency is the number of periods during 1 second:

$$\boxed{f = \frac{1}{T}} \qquad (1/s = Hz) \qquad (A26)$$

Usual frequencies in electrical power systems are:

50 Hz (e.g. Europe) 60 Hz (e.g. USA)

For 50 Hz systems, the period T is 1/50 = 20 ms.

- **ANGULAR VELOCITY OR PULSATION**

Symbol: ω **Unit: Radian per second (rad/s)**

The angular velocity or pulsation is the angle in radian covered in 1 second. It is therefore proportional to the frequency f.

$$\boxed{\omega = 2\pi f} \quad (rad/s) \qquad (A27)$$

7.3 Phasor Diagrams

Sinusoidal waveforms can be represented by phasors or vectors which might be added or subtracted like vectors of velocity diagrams. Figure A30 shows two sinusoidal waves, current i and voltage u, at frequency f and out of phase by an angle ϕ. Current and voltage can be represented in the phasor diagram as two vectors rotating at an angular velocity of $\omega = 2 * \pi * f$ and with a phase angle of $\phi = \alpha - \beta$.
α and β are the initial angles at t = O.

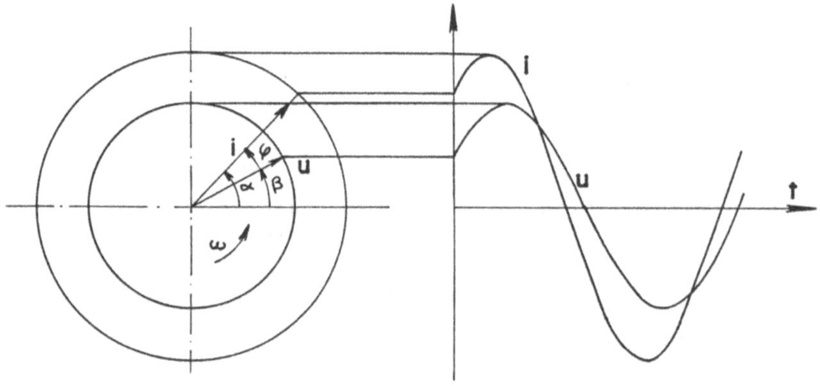

FIGURE A 30 :
Phasor diagram

7.4 Notations and Definitions

Instantaneous values are represented by lower case letters (u, i) and are the current and voltage values at a specified time t.

- **Peak values** are the maximum values of voltage or current reached during a period T and are indicated by $\hat{U}$ and $\hat{I}$.

The following relationship applies:

$$u = \hat{U} * \sin(\omega\, t + \beta)$$
$$i = \hat{I} * \sin(\omega\, t + \alpha)$$
$$\text{if } \beta = 0, \alpha = \phi \quad ==> \quad u = \hat{U} * \sin(\omega\, t) \quad \text{and } i = \hat{I} * \sin(\omega\, t + \phi)$$

- **Mean values**

In general, the mean voltage $\bar{U}$ or the mean current $\bar{I}$ of periodically alternating values are defined as follows:

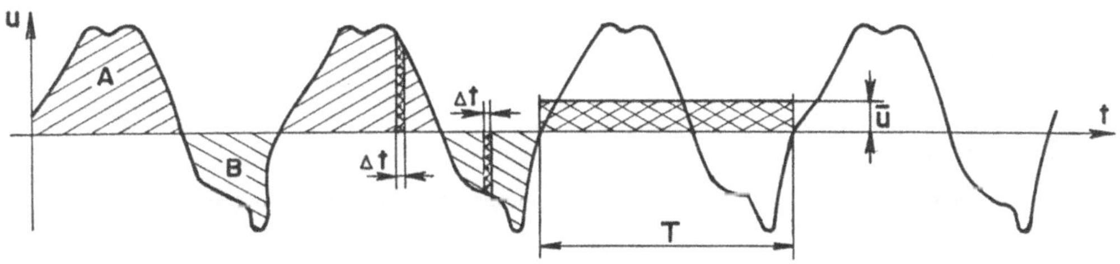

Area A - area B = $\Sigma u^+ \Delta t - \Sigma u^- \Delta t = \bar{U}\ T$ $\bar{U}$ = mean value

$$\bar{U}\ T = \Sigma u^+ \Delta t - \Sigma u^- \Delta t = \Sigma u \Delta t$$

Thus : $\bar{U} = \frac{\Sigma u\ \Delta t}{T} \qquad \bar{I} = \frac{\Sigma i\ \Delta t}{T}$

FIGURE A 31 :
Mean value of an alternating voltage

During one half wave (one alternation = T/2) the mean value is:

$$\bar{U} = \frac{2}{\pi}\hat{U} = 0.637\,\hat{U} \quad \text{and} \quad \bar{I} = 0.637\,\hat{I} \qquad \text{(A28)}$$

- Root Mean Square Values (RMS) or effective values

Peak and instantaneous values of voltage and current are not sufficient to determine the power involved in a circuit. We have to compare thermal energy produced by a direct current to the one produced by alternating current.

The RMS value I of an alternating current is that value which will produce the same amount of heat over one period as the same value of direct current.

During the time T (equal to one period), the thermal energy W_c produced by a direct current is:

$$W_c = \text{Power } P * T = R * I_c^2 * T \qquad (W*s = J)$$

In alternating current, thermal energy produced varies continuously; for a very short period dt:

$dW_a = R * i^2 * dt$ and with $i = \hat{I} * \sin(\omega t)$ the total energy produced over one period T becomes:

$$W_a = 1/2 * R * \hat{I}^2 * T$$

Per definition $W_c = W_a$. Thus:

$$R * I_c^2 * T = 1/2 * R * \hat{I}^2 * T$$

$$\boxed{I_c = I_{effective} = \frac{\hat{I}}{\sqrt{2}} = 0.707\,\hat{I}} \qquad \text{(A29)}$$

And similarly for voltage:

$$\boxed{U_c = U_{effective} = \frac{\hat{U}}{\sqrt{2}} = 0.707\,\hat{U}} \qquad \text{(A30)}$$

7.5 Power in AC

The instantaneous power is the product of the instantaneous values of voltage and current:

$$p(t) = u(t) * i(t)$$

Using the definitions above:

$$p(t) = \hat{U}*\sin(\omega t + \beta) * \hat{I}*\sin(\omega t + \alpha)$$

After goniometric transformation:

$$p(t) = 1/2*\hat{U}*\hat{I}*\{\cos(\alpha + \beta) - \cos(2\omega t + \alpha + \beta)\}$$

Replacing the peak values $\hat{U}$ and $\hat{I}$ by the effective values:

$$p(t) = U * I * \{\cos(\alpha + \beta) - \cos(2\omega t + \alpha + \beta)\}$$

Putting $\alpha - \beta = \phi$ and if $\beta = 0$ then $\alpha = \phi$:

$$\boxed{p(t) = U\,I\,\{\cos\phi - \cos(2\omega t + \phi)\}} \qquad \text{(A31)}$$

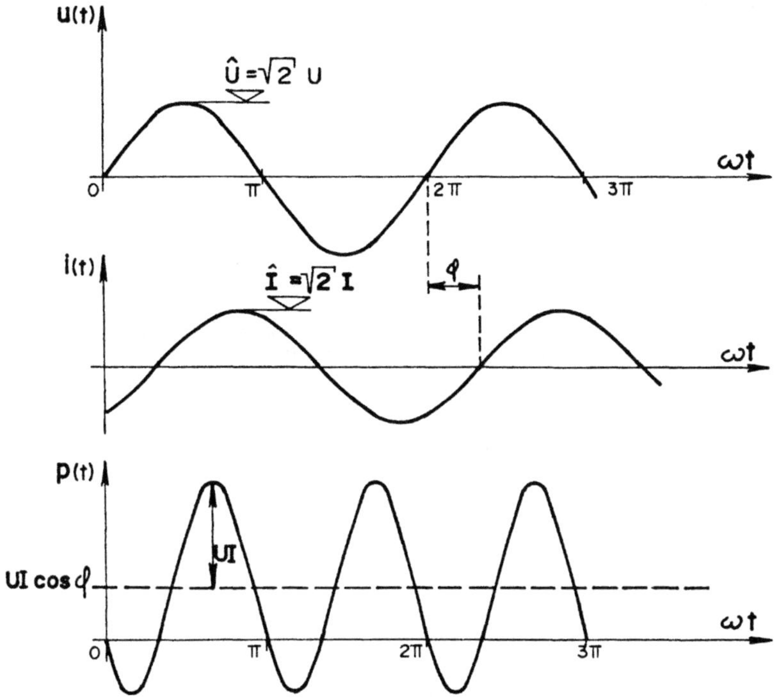

FIGURE A 32 : Voltage, current and power variations versus time

Plotting instantaneous power versus time as in Figure A 32, we can observe two main features of AC systems:

- instantaneous power in AC pulsates at double the frequency of the grid voltage (this explains why lamps pulsate at 100 Hz in a 50 Hz grid; however, the eye cannot distinguish this frequency from any other down to about 18 Hz);

- instantaneous power comprises two terms (from equation above):

a constant power $P_a = U*I*\cos\phi$ and

a variable power $p_b = - U*I*\cos(2\omega t + \phi)$

Three distinct values of ϕ are further examined:

a) current and voltage are in phase (ϕ = 0); thus P becomes P_a because the mean value of the second term is zero. This occurs for a circuit connecting pure resistors and we can write:

$P = U * I$ which is the same as in DC circuits
(note U and I are RMS or effective values of the alternating current and voltage).

In more general terms, this is called

- **ACTIVE POWER**

Symbol: P **Unit: Watt (W)**

$$P = U\ I\ \cos\phi \qquad \text{(A32)}$$

where the example above with the pure resistive load is only a special case, namely the maximum active power obtainable (cos (0) = 1).
The active power corresponds to a real energy conversion from electricity into heat or work. When measuring power with a wattmeter, it is the active power which will be displayed on the meter.

b) current and voltage are 90° out of phase (ϕ = 90°). Plotting instantaneous power against time we obtain positive and negative values with a mean value of zero. This is called reactive power.

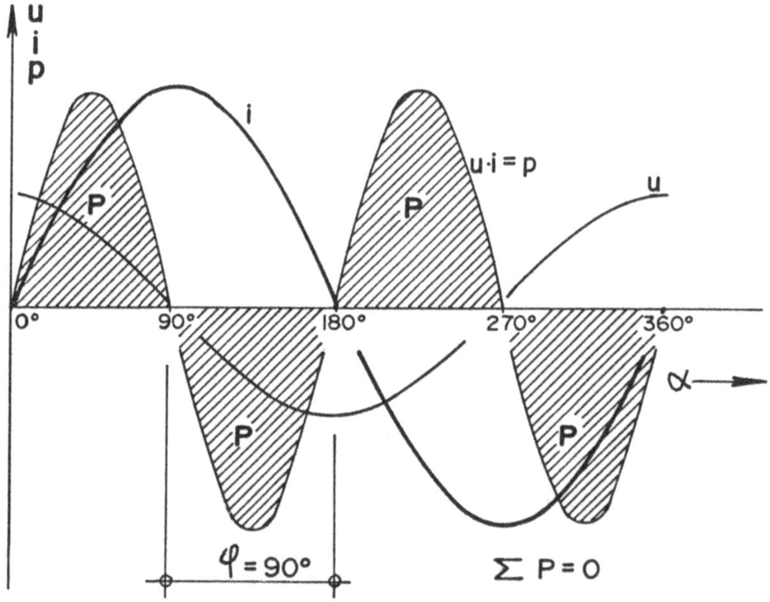

FIGURE A 33 :
Pure Reactive Power

- REACTIVE POWER

Symbol: Q **Unit: Reactive volt-ampere (VAr)**

Definition:

$$\boxed{Q = U\ I\ \sin\phi} \qquad (A33)$$

where our example with $\phi = 90^o$ corresponds to the special case of maximum reactive power (sin 90° = 1). For $\phi = 90^o$ no active power occurs. Power only oscillates in the circuit and is used to build up magnetic and electric fields in inductances and capacitors.

Referring to Figure A33, we can see that between t = 0 and $\pi/2$ a magnetic field (in a coil for example) builds up, the coil consumes energy from the source while during the next half-wave the magnetic field disappears, the coil delivers energy back to the source. Thus, no useful energy occurs.

c) phase angle ϕ between voltage and current is somewhere between 0 and 90°. This is a circuit comprising coils or capacitors plus resistors. In such a circuit the reactive power which builds up magnetic and electric fields and which was said to correspond to no thermal or mechanical energy, creates losses. It can be imagined that a certain amount of power is required to force the electrons oscillating between source and capacitors/inductances through the resistors of the circuit (conductors). So, reactive power brings about losses in resistors in the form of heat as does active power. Therefore, total power to be delivered by the source must be higher than the useful mechanical or thermal power (active power) required by the loads of the circuit. This total power is called apparent power. All the equipment of a network (generator, conductors) must be designed to deliver and transmit the apparent power.

- APPARENT POWER

Symbol: S **Unit: Volt-ampere (VA)**

By definition:

$$\boxed{S = U\ I} \qquad (A34)$$

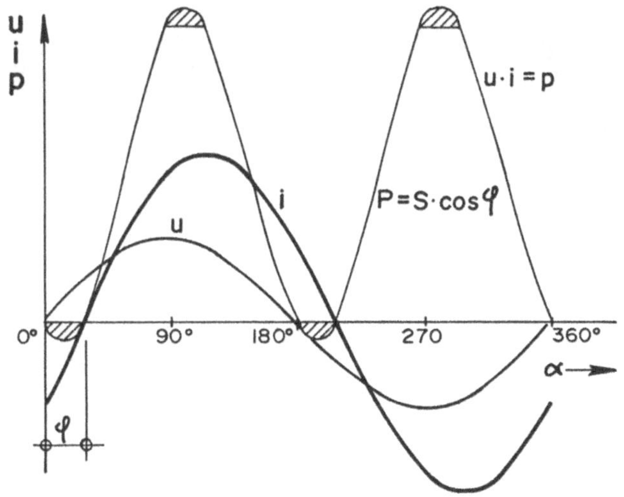

FIGURE A 34 :
Apparent power

The three forms of power (active, reactive and apparent) are linked together in the so-called power triangle where the angle phi corresponds to the phase angle between voltage and current.

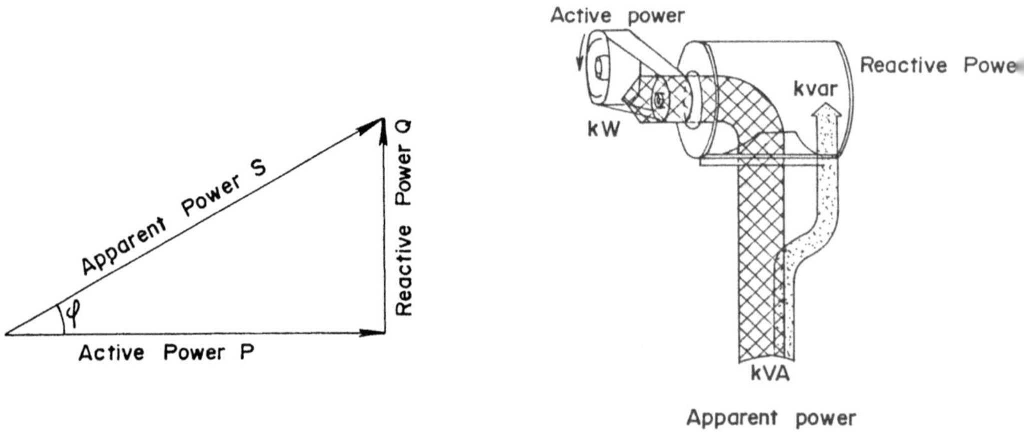

FIGURE A 35 :
Power triangle

7.6 Power factor, cos ϕ

The power factor cos ϕ is the ratio of the active power P and the apparent power S:

$$\cos\phi = \frac{P}{S} \tag{A35}$$

A mechanical equivalent of the power triangle and the concept of cos ϕ is given in Figure A36.

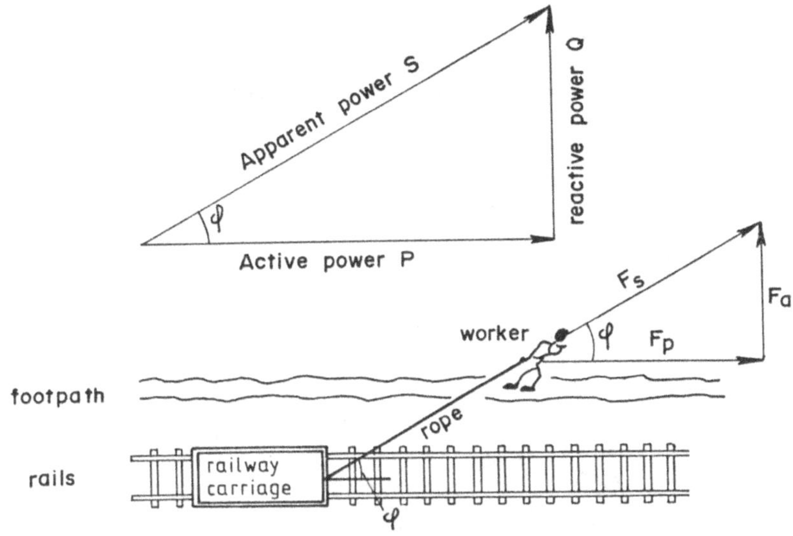

FIGURE A 36 :
Mechanical equivalent of the power triangle and cos ϕ

The force contributing to the displacement of the railway carriage is F_p. With an increasing angle ϕ tension in the rope increases but without contributing to the traction of the carriage. The force acting perpendicular to the railway lines is expended in pressing the wheels against the rails causing only increased friction losses.
A similar situation occurs in electrical systems. The larger the phase angle between current and voltage, the stronger the electrical equipment required to generate and transmit the same active power. The power factor cos phi characterizes the nature of an electrical circuit. A low power factor (high phase angle) is created by inductances (creating the magnetic field in a motor for example) connected to the system. Electricity boards try to improve the power factor by installing capacitors into the system (see section 8).

7.7 Circuit Elements in AC

7.7.1 Pure Resistive Circuit

Consider a resistance R in an AC circuit with RMS values of voltage U and current I.

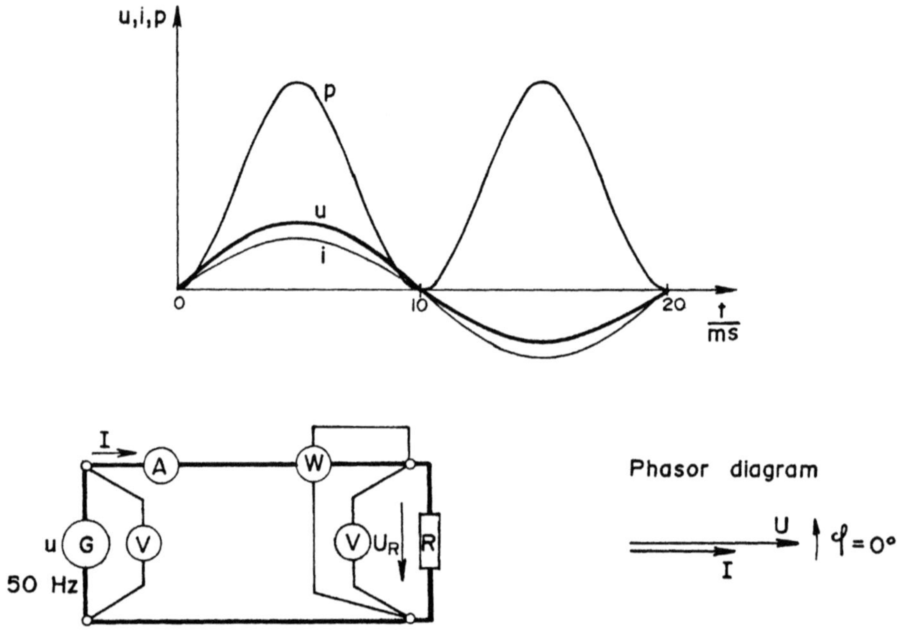

FIGURE A 37 :
Pure resistive AC circuit and instantaneous values of u, i and p

Voltage U and current I are in phase ($\phi = 0$, $\cos\phi = 1$)

Ohm's law applies in the same way as in DC circuits:

$$\boxed{U = R\ I} \qquad \text{(A36)}$$

according to the definitions given above:

$P = U{*}I{*}\cos\phi = U{*}I = S$ and $Q = S * \sin\phi = 0$

Thus:

$$\boxed{P = R\ I^2 = \frac{U^2}{R}} \qquad \text{(A37)}$$

7.7.2 Pure Inductive Circuit

Voltage of an RMS-value U is applied across a pure inductance L as shown in Figure A38.

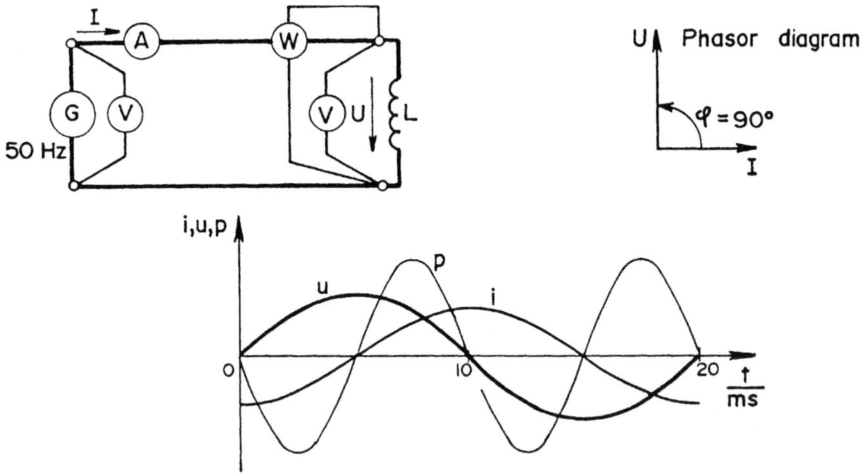

FIGURE A 38 :
AC circuit with a pure inductance

Current lags voltage by a phase angle of +90° or +π/2 radians, i.e. voltage reaches its peak before the current. This is due to the opposing effect of self-induction in a coil which prevents the current to flow in phase with the voltage. Current is thus limited by the so-called

Inductive Reactance X_L. It represents the resistance of an inductor in AC. Reactance has therefore the same unit as resistance, namely Ω . The inductive reactance is the product of the angular velocity ω and the inductance L (see section 5.1 above).

$$\boxed{X_L = \omega L = 2 \pi f L} \qquad \text{(A38)}$$

where f = frequency of voltage and current

Similar to Ohm's law:

$$\boxed{U = X_L I = \omega L I} \qquad \text{(A39)}$$

if f = 0 (ω = 0) which is DC: U becomes zero which is a short circuit.
if f = infinite then I = 0, i.e. the circuit is opened

As discussed earlier, instantaneous power fluctuates around the mean value zero. During a quarter of a period the power is positive and energy is stored in the magnetic field of the inductor and during the next quarter this energy is returned to the source. The active power is thus zero (ϕ = 90° then cos ϕ = 0 and sin ϕ = 1):

$$\boxed{P = U I \cos\phi = 0} \qquad \text{(A40)}$$

On the other hand reactive power is:

Q = U * I * sinφ = U * I and according to the definition above:

$$Q = X_L I^2 = \omega L \; I^2 = \frac{U^2}{X_L} \quad \text{(VAr)} \qquad \text{(A41)}$$

Inductances consume reactive power

7.7.3 Pure Capacitive Circuit

Voltage of an RMS- value U is applied across a pure capacitance C as shown in Figure A39.

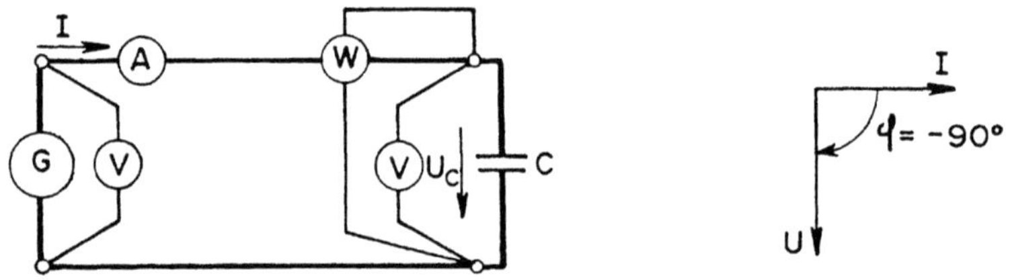

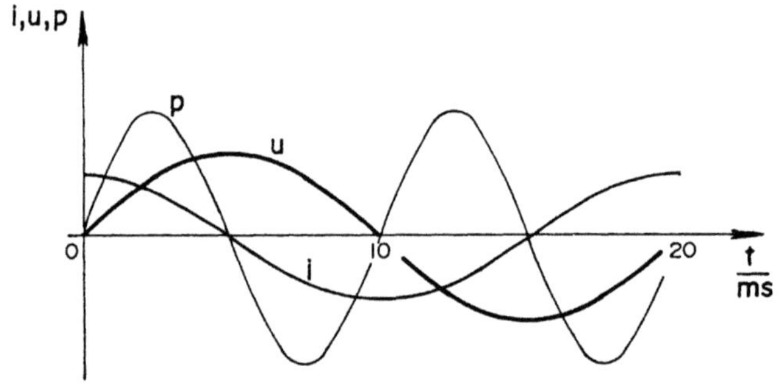

FIGURE A 39 :
AC circuit with a pure capacitance

Voltage lags current by a phase angle of -90° or -π/2 radians, i.e. the current reaches its peak before the voltage. This is due to the fact that the capacitor is to be charged first by electrons before peak voltage across the capacitor is built up. Voltage is thus limited by the so-called

Capacitive Reactance X_C. It represents the resistance of a capacitance in AC. Reactance has therefore the same unit as resistance, namely Ω . The capacitive reactance is the reciprocal value of the product of the angular velocity ω and the capacitance C (see section 2.5.2 above).

$$\boxed{X_C = \frac{1}{\omega\ C} = \frac{1}{2\pi f C}} \qquad (A42)$$

Similar to Ohm's law:

$$\boxed{U = X_C I = \frac{I}{\omega\ C}} \qquad (A43)$$

if f = 0 (ω = 0) which is DC: I becomes zero, i.e. the capacitor blocks a direct current.
if f = infinite then U = 0, which is a short circuit

As discussed earlier, instantaneous power fluctuates around the mean value zero. During a quarter of a period the power is positive and energy is stored in the electrostatic field of the capacitor and during the next quarter this energy is returned to the source. The active power is thus zero (ϕ = -90^0 then cos ϕ = 0 and sin ϕ = -1):

P = U * I * cos ϕ = 0

On the other hand, reactive power is:

Q = U * I * sin ϕ = U * I and according to the definition above:

$$\boxed{Q = X_C\ I = \frac{I^2}{\omega\ C} = \frac{U^2}{X_C}} \quad \text{(VAr)} \qquad (A44)$$

Capacitance generates reactive power

7.8 Impedance Z in Series Connections

Consider a resistance and an inductance in series in an AC circuit (see Figure A40). In the resistance, voltage and current are in phase while in the inductance, voltage leads current by 90°. However, the same alternating current flows through both elements, which creates a voltage drop in both. In AC one has to consider that the two voltage drops are not in phase. A resulting voltage sine-wave has to be constructed using vector addition. The most convenient way to do so is the phasor diagram presented in section 7.3; as magnitudes of the phasors, the RMS values are used rather than instantaneous values of voltage (and current).

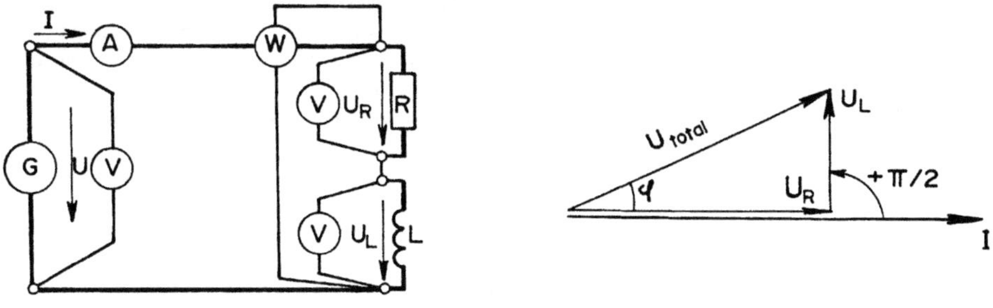

FIGURE A 40 :
AC circuit with a resistance and an inductance in series and phasor diagram of voltage

The individual voltage drops of the elements are:

$U_R = R * I$ and $U_L = X_L I$
the resulting voltage is

$$U = \sqrt{U_R^2 + U_L^2} = I \sqrt{R^2 + X_L^2} \qquad (A45)$$

the square root is called **Impedance Z**

$$Z = \sqrt{R^2 + X_L^2} = \frac{U}{I} \qquad (A46)$$

(compare with Ohm's law $R = U/I$)

Similarly, if a capacitance is added in the circuit of Figure A40 the resulting voltage is composed of three individual voltage drops with the additional

$U_C = X_C * I$ which lags the current by 90^0 (which is 180^0 from U_L).
Thus, the general formula of impedance in series connection becomes:

$$Z = \sqrt{R^2 + (X_L - X_C)^2} = \sqrt{R^2 + (\omega L - \frac{1}{\omega C})^2} \qquad (A47)$$

(impedance for series connection of resistance, inductance and capacitance in AC)

The phase angle between the current I and the resulting voltage U is:

$$\tan \phi = \frac{X_L - X_C}{R} \qquad (A48)$$

7.9 Impedance Z in Parallel Connections

Consider an AC circuit with a resistance, an inductance and a capacitance in parallel (see Figure A41). From DC circuits we know that the voltage across each element is the same as the source voltage U. The current through the resistor is obviously in phase with the source voltage whereas the currents of the inductance and the capacitance lag or lead it by 90° or -90° respectively.
To compute the resulting current, a phasor diagram of the current vectors can be constructed (again using the RMS values of the individual currents).

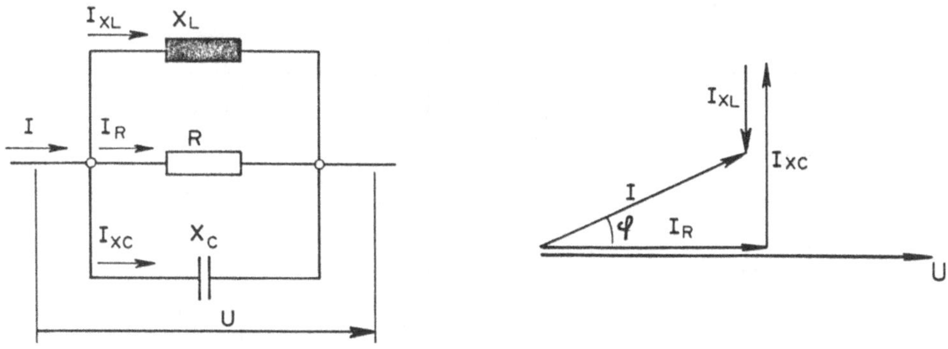

FIGURE A 41 :
AC circuit with different elements in parallel and phasor diagram of currents

From the phasor diagram we obtain immediately the resulting current:

$$I = \sqrt{I_R^2 + (I_L - I_C)^2} \qquad \text{(A49)}$$

The individual currents of the elements are:

$\mathbf{I_R = U / R}$ (in phase with U)
$\mathbf{I_L = U / X_L = U / (\omega L)}$ (out of phase with U by +90°)
$\mathbf{I_C = U / X_C = U \ \omega \ C}$ (out of phase with U by -90°)

Impedance Z was defined as Z = U / I hence:

$$Z = \sqrt{\frac{1}{\frac{1}{R^2} + (\frac{1}{X_L} - \frac{1}{X_C})^2}} \qquad \text{(A50)}$$

(impedance for parallel connection of resistance, inductance and capacitance in AC)

The phase angle between the resulting current I and the voltage U is:

$$\tan \phi = \frac{R}{X_L - X_C} \qquad \text{(A51)}$$

The following table gives an overview of all possible connections in AC with relevant formulae for the resulting current/voltage and power.

TABLE A 42 :
Overview of connections of elements in AC and relevant formulae

<table>
<tr><th>Connection</th><th>Phasor diagram</th><th>cos ϕ</th><th>Power</th></tr>
<tr><td><u>1. Pure Elements:</u>
a) resistance R
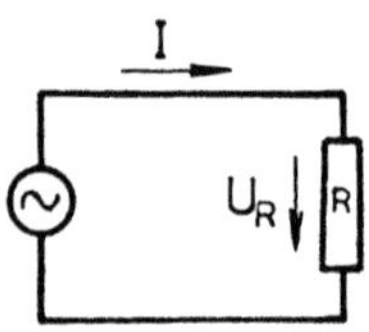</td><td>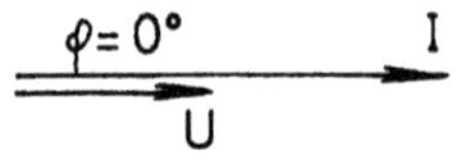
$U = R\,I$</td><td>$\cos\phi = 1$</td><td>$P = R\,I^2 = \frac{U^2}{R}$
$S = P$
$Q = 0$</td></tr>
<tr><td>b) inductance L
</td><td>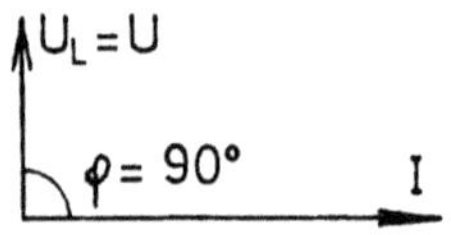
$U_L = X_L\,I = \omega L I$</td><td>$\cos\phi = 0$</td><td>$P = 0$
$S = U\,I$
$Q = X_L\,I^2 = \frac{U2}{X_L}$</td></tr>
<tr><td>c) capacitance C
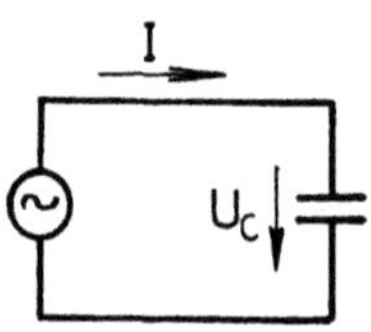</td><td>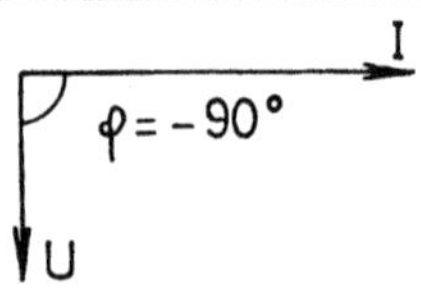
$U_C = X_C\,I = \frac{I}{\omega\ C}$</td><td>$\cos\phi = 0$</td><td>$P = 0$
$S = U\,I$
$Q = X_C\,I^2 = \frac{U^2}{X_C} = \frac{U^2}{\omega\ C}$</td></tr>
<tr><td><u>2.Series Connection</u>
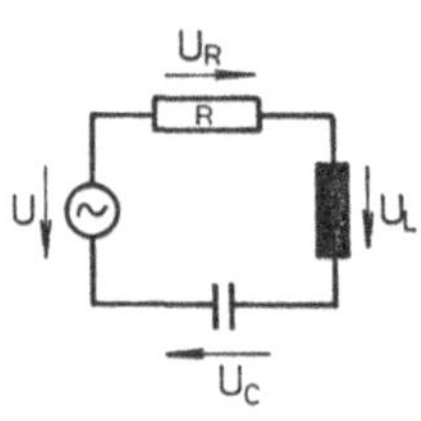</td><td>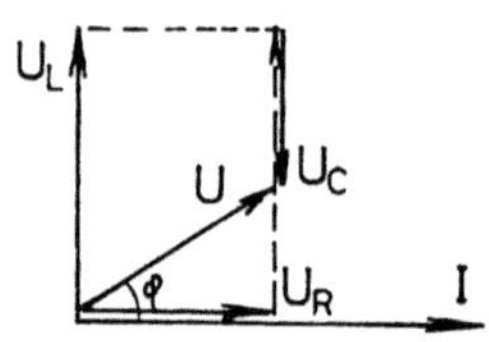
$U = Z\ I$ where
Z = impedance
$Z = \sqrt{R^2 + (X_L - X_C)^2}$</td><td>$\tan\phi = \frac{X_L - X_C}{R}$</td><td>$P = R\,I^2 = U^2/R$
$S = U^2 / Z = Z\,I^2$
$Q = (X_L - X_C)\,I^2$</td></tr>
<tr><td><u>3.Parallel Connection</u>
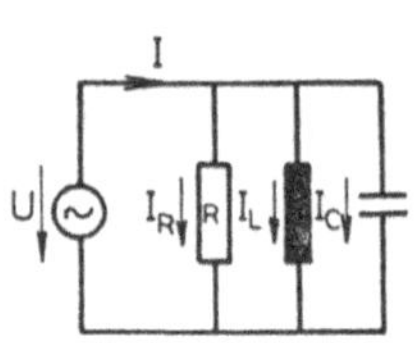</td><td>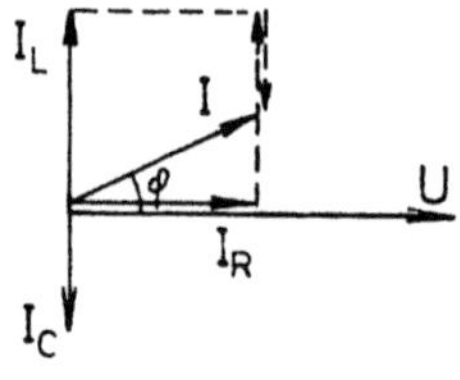
$I = U/Z$
$Z = \sqrt{\frac{1}{1/R^2 + (1/X_L - 1/X_C)^2}}$</td><td>$\tan\phi = \frac{R}{X_L - X_C}$</td><td>$P = R\,I^2 = U^2/R$
$S = U^2/Z = Z\,I^2$
$Q = \frac{U^2}{X_L - X_C}$</td></tr>
</table>

7.10 Resonance

7.10.1 Resonance for series connection of L and C

Consider a circuit with a resistance R, an inductance L and a capacitance C connected in series. Current is determined by

$$I = \frac{U}{Z} = \frac{U}{\sqrt{R^2 + (X_L - X_C)^2}} \qquad \text{(A52)}$$

and reaches a maximum value when the impedance is minimum. This happens if $X_L = X_C$ and is known as resonance. Since reactance is zero, current depends only on the resistor; thus $Z = R$ and $I = U/R$. The phase angle between current and voltage is eliminated.

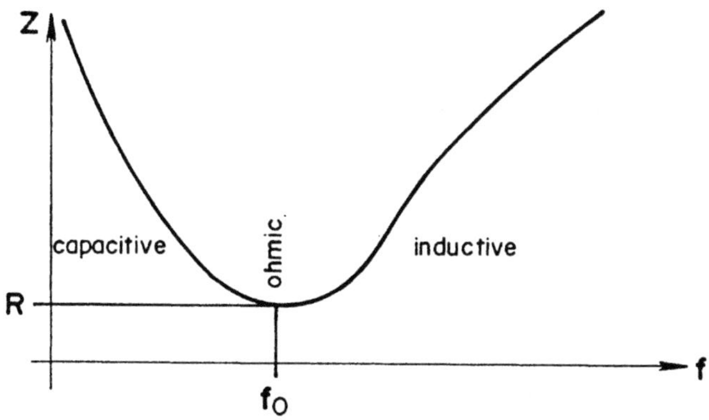

FIGURE A 43 :
Varying impedance as a function of frequency and resonance frequency for series connection

In practice, it is important to know at what frequency resonance for a given circuit occurs. The condition is $X_L = X_C$ and we obtain:

$$\omega_o L = \frac{1}{\omega_o C} \Longrightarrow \omega_o^2 LC = 1$$

$$\boxed{f_o = \frac{1}{2\pi \sqrt{LC}}} \qquad \text{(A53)}$$

Resonance in series connections can become dangerous because the voltage across the inductance and capacitance reaches very high values. Due to their opposite position (see phasor diagram) these voltage values equal each other out and do not have any effect on the circuit but the coil and the capacitor themselves could be damaged.

7.10.2 Resonance for parallel connection of L and C

In the same way as for series connection, resonance can occur also for parallel connection of inductances and capacitors. In this case it is a resonance of currents, i.e. the currents flowing through the inductor and the conductor equal each other out at a certain frequency, the resonance frequency which takes the same form as with series connection (see Formula A53). But contrary to that case, the current of the circuit takes a **minimum** value at the resonance frequency ($Z = R$).

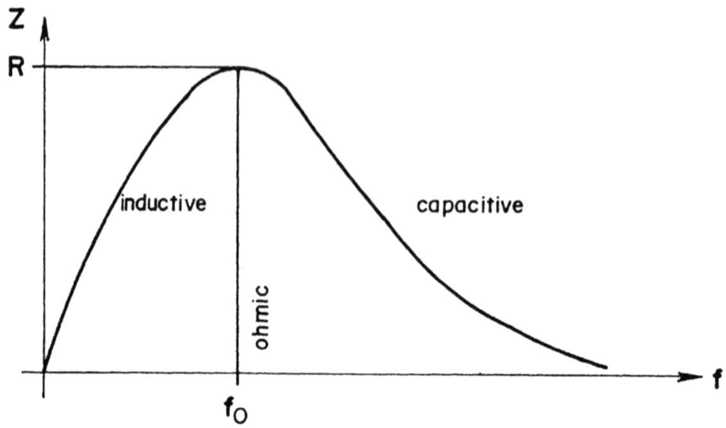

FIGURE A 44 :
Varying impedance as a function of frequency and resonance frequency for parallel connection

7.11 Correction of the Power Factor cos phi

7.11.1 General

The previous sections have shown how the phase angle and thus reactive power for a certain configuration of inductances and capacitors can be reduced or eliminated. The same principle is used to improve the power factor of devices requiring reactive power.

The importance of a small phase angle between voltage and current (i.e. power factor cos ϕ close to unity) can be summarized as follows: for any power factor deviating from unity, power and therefore current to be generated and distributed by the electricity board is higher than the active power and current which can be converted into useful work in customers' appliances. Less power to be distributed for the same useful work means smaller conductors and generators and thus less costly installations.

In an inductive circuit, power factors are improved by connecting capacitors which generate reactive power at the device, thus reducing the reactive power demand from the grid. Capacitors are connected:

- in parallel with motors;
- in series with fluorescent tubes.

Note that the inductance of a device remains the same; i.e. its cos ϕ does not change but the power factor of the whole installation is improved. Reactive power oscillates between capacitor and device and, therefore, does not put load on the grid.

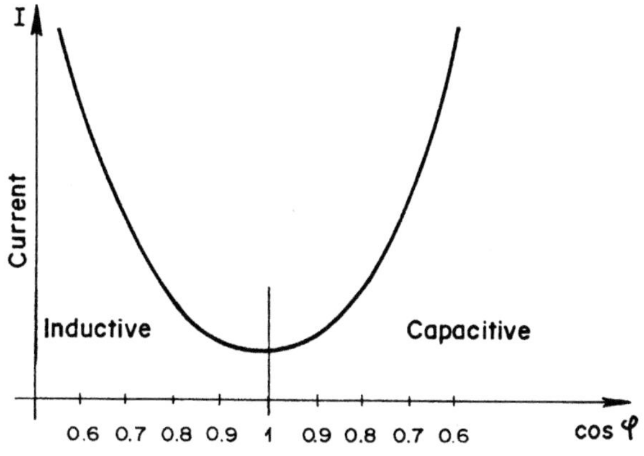

FIGURE A 45 :
Variation of grid current as a function of cos ϕ

If cos $\phi = 1$, then grid current I reaches a minimum value

7.11.2 Calculation of the Compensating Capacitance

Consider a single-phase network having a rated voltage U and supplying a motor. The power factor should be corrected from cos ϕ_1 to cos ϕ_2 by capacitors while the active power P remains the same.

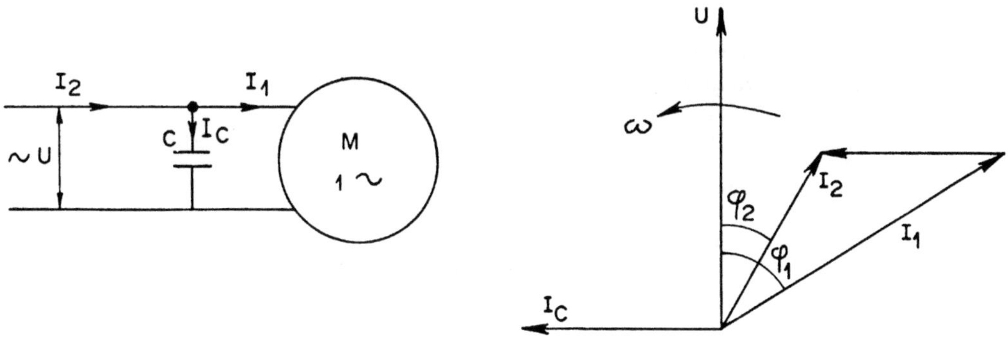

FIGURE A 46 :
Power factor correction of a motor

- Active power $P = U * I_1 * \cos \phi_1 = U * I_2 * \cos \phi_2$ where

  I_1 = grid current before compensation
  I_2 = grid current after compensation

- Reactive power Q:

- before compensation $Q_1 = P * \tan \phi_1$
- after compensation $Q_2 = P * \tan \phi_2$

The reactive power to be generated by the capacitance is:

$$Q_C = Q_1 - Q_2 = U^2 / X_c = U^2 \omega\ C = U^2\, 2\pi\, f\, C$$

(see Table A 42 above)

$$\boxed{C = \frac{Q_1 - Q_2}{U^2\ 2\pi\ f} = \frac{P\,(\tan\phi_1\ -\ \tan\phi_2)}{U^2\ 2\pi\ f}}\ \text{(Farad)} \qquad \text{(A54)}$$

The new apparent power to be supplied by the network reduces to:

$$S_2 = P / \cos \phi_2 \qquad \text{(VA)}$$

The new current to be supplied by the grid reduces to:

$$I_2 = S_2 / U \qquad \text{(A)}$$

Example 4:
A workshop has the following electrical devices:

- Lamps, electric stove: (pure resistive loads) Total 5 KW

- Electric motor A: rated output 5 kW, efficiency 0.8, $\cos \phi = 0.75$
- Electric motor B: rated output 1.5 kW, efficiency 0.75, $\cos \phi = 0.7$

The workshop is to be connected to the local grid:
single phase, rated voltage 220 V, 50 Hz.
Power factor required min. 0.9 for each motor

Calculate the capacitors required for the two motors (parallel connection). Determine also the cross sectional area of the supply lines if the distance to the grid is 500 m and the voltage drop should not exceed 15 V (using copper wire, mean temperature 20° C).

What is the total power demand (active and apparent) of the workshop assuming simultaneous operation of all loads? What will be the power factor for this case?

8. THREE-PHASE SYSTEM

8.1 General

In Chapter 7, the AC system has been introduced on the basis of a single-phase network. In practice, the single-phase system is only used for small networks (< 5 kW). Transmission of bulk power over long distances is more economical using the three-phase system.

Symbol: 3 ~

Conveying the same power using a three-phase system instead of a single-phase one requires less than a third of conductor material (cross sectional area) despite the fact that a greater number of conductors (4 instead of 2) are used. Additionally, three-phase systems provide two different voltage levels depending on how the end-use appliances are connected to the grid (see below).
The disadvantage of the three-phase system is its complexity (switchgear, monitoring, control and protection equipment for three phases rather than for one only) and the need to balance the loads over the three phases.

8.2 Generating Three-Phase Voltage and Current

Three-phase generators may be described as 3 single-phase machines rotating on the same shaft whereby the coils receiving induced voltage are all displaced to each other by 120°. Figure A47 below shows a schematic diagram of a two-pole AC generator for a) a single phase system and the generated voltage while b) provides the corresponding three-phase system. The phasor diagram drawn from the rms voltage values of each phase shows the phase angles of 120° between the three induced voltages (or currents).

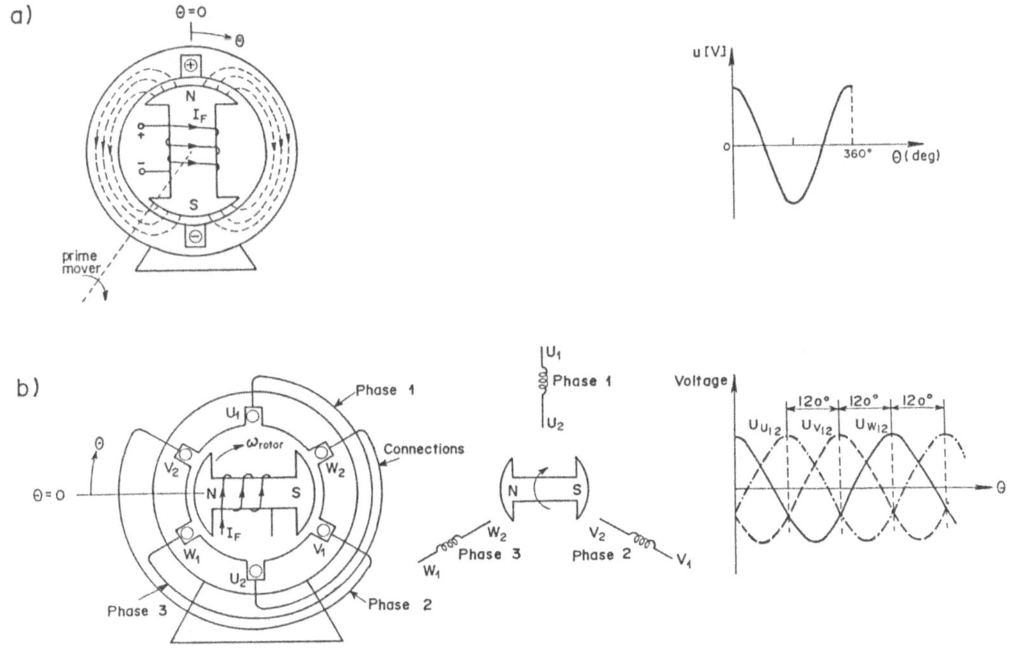

FIGURE A 47 :
Two-pole AC generator for a) single phase network and b) three phase systems and the generated voltage

Note that the construction of the AC generator is fundamentally different from the DC generators or motors as described above although AC voltage and current may be obtained from a DC machine simply by replacing the commutator by slip rings. In practice, this is never done; AC machines employ a rotating magnetic field to induce voltage in stationary coils. Thus, the role of the stationary and rotating member is interchanged as compared to the DC machine (more details on the design of AC machines see main text).

Examining the waveforms of the three voltages produced or the phasor diagram, we arrive at the following conclusion:

The sum of the three voltages or currents at any instant (t) is zero (valid for balanced, i.e symmetrically loaded three-phase systems).

Figure A48 shows this graphically on the basis of the phasor diagram.

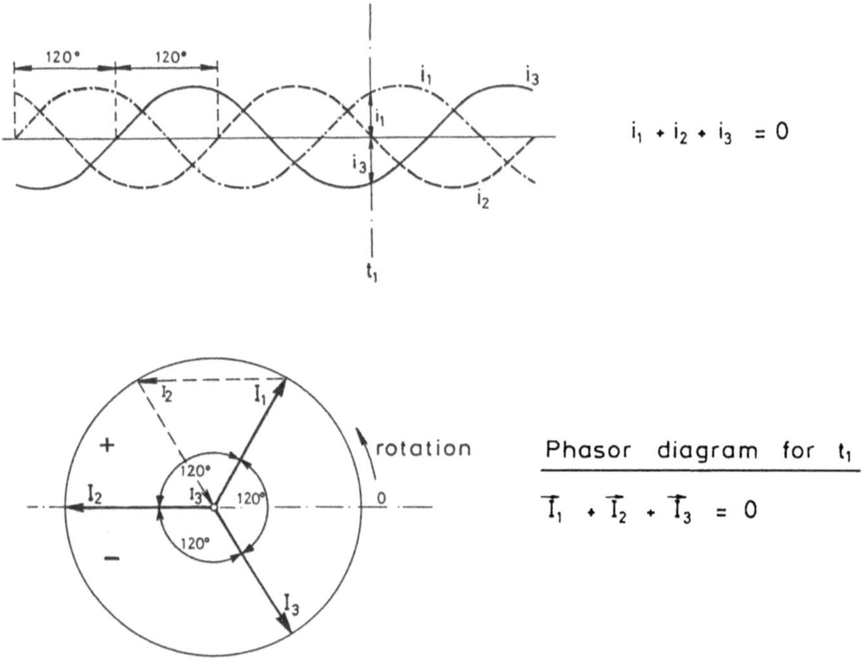

FIGURE A 48 :
The sum of the voltage values in the three-phase system

8.3 Connections

Instead of using 6 conductors (input and output of each coil) to transmit electricity in a three-phase system from the generator to the end-users, two or three conductors can be omitted. The three coils of the stator are interconnected either in star (or wye (Y)) or in delta (Δ) to produce a three phase voltage source as will be illustrated below.

8.3.1 Definitions

<u>**Phase Voltage**</u>
This is the voltage between the terminals of a coil or between any coil terminal and the neutral.

<u>**Line Voltage**</u>
This is the voltage appearing between any of two conductors.

<u>**Phase current**</u>
This is the current flowing through the stator windings/coils of the generator (or motor).

Line current
This is the current flowing through the conductors between the power station and the end-users.

8.3.2 Star or Wye Connection

Symbol : **Y**

One end of all three stator coils are interconnected to form the letter wye. The central point is called the neutral and may be brought out to end-users and might be used for earthing.

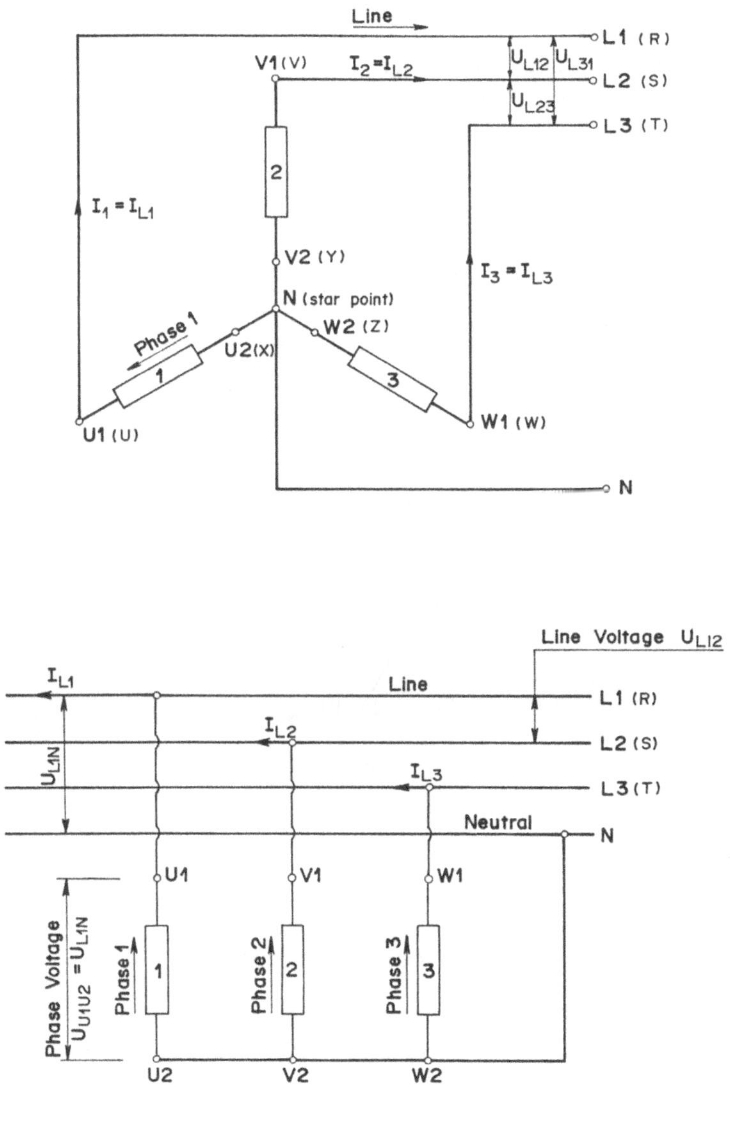

FIGURE A 49 :
Three-phase voltage source Y connected
(old designations in brackets)

From such a system, the following voltages and currents are available:

$$\text{Phase Voltage } U_{ph1} = U_{L1\,N} = U_{U1\,U2}$$
$$\text{Phase Voltage } U_{ph2} = U_{L2\,N} = U_{V1\,V2}$$
$$\text{Phase Voltage } U_{ph3} = U_{L3\,N} = U_{W1\,W2} \qquad \text{(A55)}$$

$$\text{Line Voltages } U_{line} = U_{L1\,L2} = U_{U1V1} \quad , \quad U_{L2\,L3} = U_{V1W1} \quad , \quad U_{L3\,L1} = U_{W1U1} \qquad \text{(A56)}$$

The magnitude of the line voltgage may be obtained graphically or analytically from the phasor diagram by subtracting the first phase voltage from the second (from Figure 50, schematic layout on the left: $\Sigma U = 0$ within one mesh: $\underline{U}_{L1\,L2} = \underline{U}_{L1\,N} - \underline{U}_{L2\,N}$)

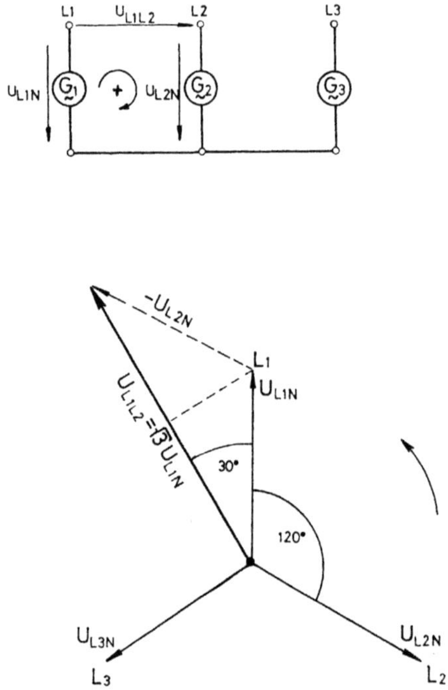

FIGURE A 50 :
Development of the line voltage in Y-connection

We find:

$$U_{line} = U_{L1\,L2} = \sqrt{3}\, U_{L1\,N} \quad \text{or } U_{line} = \sqrt{3}\, U_{phase} \quad \text{in Y - connection} \qquad \text{(A58)}$$

This is one of the advantages of the Y - connection: two different voltage levels may be obtained from the same distribution network.

In Y - connection, the phase current I_{ph} is equal to the line current I_{linie}

$$I_{line} = I_{phase} \quad , \quad I_{L1} = I_1 \quad \text{for Y - connection} \qquad \text{(A59)}$$

8.3.3 Delta Connection

Symbol: Δ

This is another possibility to save on conductors: In delta connection, the output terminals of each coil are connected to the input terminals of the following coil.

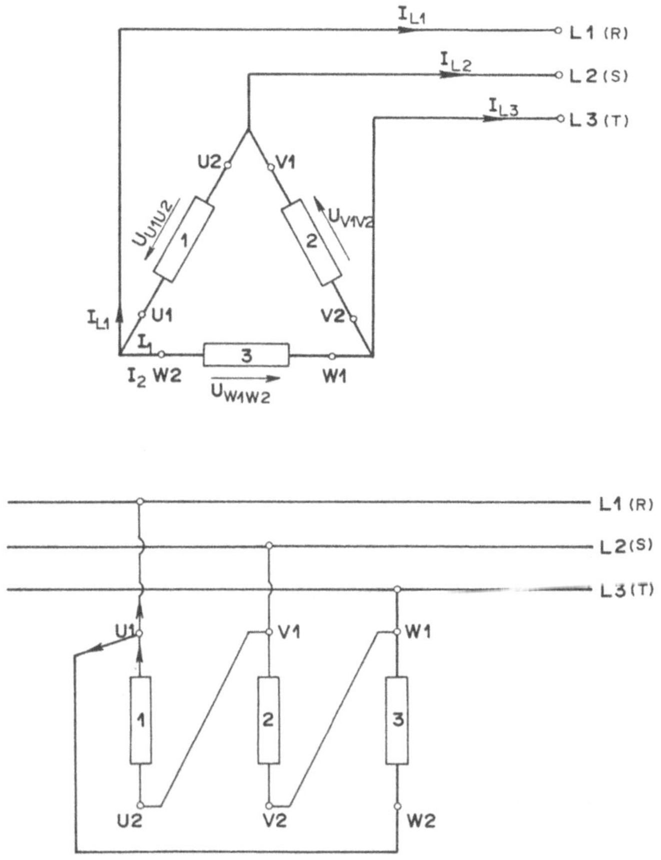

FIGURE A 51 :
Three-phase voltage source delta connected

<u>**Voltage**</u>

The line conductors are directly interconnected via the stator windings and we find:

$$\boxed{U_{line} = U_{L1\,L2} = U_{U1U2} = U_{phase}} \text{ for } \Delta \text{ connection} \qquad \text{(A60)}$$

<u>**Current**</u>

At each terminal of the coils (corner points of the triangle) three currents branch off: the phase current I_1 flows towards the terminal, phase current I_3 flows into the coil while the difference of these two induced currents must obviously flow into the line (conductor) L1. Thus:

$$\underline{I}_{L1} = \underline{I}_1 - \underline{I}_3$$

This equation can be solved graphically using the phasor diagram for currents.

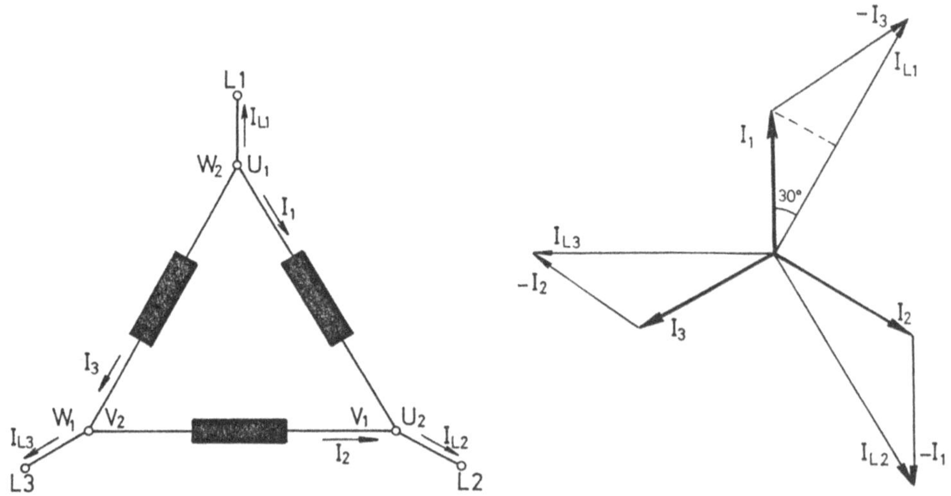

FIGURE A 52 :
Development of line currents in delta connection

We find:

$$\boxed{I_{L1} = \sqrt{3}\, I_1 \quad \text{or,} \quad I_{\text{line}} = \sqrt{3}\; I_{\text{phase}}} \quad \text{for } \Delta \text{ connection} \qquad \text{(A61)}$$

8.3.4 Power in three-phase systems

Since a three-phase system is equivalent to 3 single-phase systems, we can write:

Apparent power $\mathbf{S = 3\, U_{phase}\, I_{phase}}$

Expressing phase current and voltage by the relationships given above for star and delta connections we find for <u>both</u> types of connections:

$$\boxed{S = \sqrt{3}\;\; U_{\text{line}}\, I_{\text{line}}}\ \text{(VA)} \quad \text{for Y and } \Delta \text{ connection} \qquad \text{(A62)}$$

<u>Impedance</u>

We define an impedance of phase Z, which is the ratio between voltage and current:

$\mathbf{Z_{phase} = U_{phase} / I_{phase}}$

$$\text{Y connection:} \quad Z_{Yphase} = \frac{U_{line}}{\sqrt{3}\, I_{line}} \qquad \text{(A63)}$$

$$\Delta \text{ connection:} \quad Z_{\Delta phase} = \frac{U_{line}\sqrt{3}}{I_{line}} \qquad \text{(A64)}$$

8.3.5 Conclusion

Y connection gives a higher terminal voltage than delta connection but a correspondingly smaller output current.

The same relationships are valid not only for generators but also for motors and other end-use appliances. Using the same principle, a motor of a certain impedance connected in star (Y) to a given network voltage (U_{line}) will draw only a third of the current as compared to delta connection (Δ). This feature is used to limit the starting current of asynchronous motors which usually have starting currents of 4.5 to 6 times the nominal current; i.e. the motor is first connected in Y up to almost nominal speed, whereby drawing a current only slightly above nominal (1.5 to 2 times), and then switched into Δ connection to develop full operating power and torque.

Comparing impedances in the Y and Δ connection, we may conclude the following (see Formulae A63 and A64 above):

To arrive at the same voltage and current levels, phase impedance in Δ connection must be three times higher than the corresponding impedance in Y connection .

$$Z_{\Delta} = 3\ Z_Y \qquad \text{(A65)}$$

8.4 Rotor Speed and Generated Frequency

Rotor speed and frequency of the generated voltage are directly related. Consider the elementary two-pole AC generator of Figure A47 above. Applying DC current to the field winding on the rotating rotor creates a rotating magnetic field which induces voltage in the stator coil(s).
Obviously, turning the rotor one complete turn generates one cycle of AC voltage. Rotating the rotor one turn in one second gives one cycle per second or 1 hertz (Hz) as the frequency is referred to (frequency is directly proportional to the speed of the rotor; see also section 7.2 above).
If the rotor has more than two poles, say 2p poles, each revolution of the rotor induces p cycles of voltage in the stator coils. Note that due to the continuous nature of the magnetic field lines, the magnetic poles always occur in pairs, hence, 2p can only take the values of 2, 4, 6 etc. Thus, we can write in accordance with Formula (A27) above:

$$f = p\frac{n}{60} \quad \text{(Hz)} \qquad \text{(A66)}$$

where f = generated frequency (1/s = Hz)
n = rotor speed (rpm, revs per minute)
p = number of pole pairs (where p must always be an integer)

To produce voltage of a given frequency, the rotor must be turning at a speed compatible with this frequency, i.e. the rotor must rotate at **synchronous speed**. The common frequencies of 50 and 60 Hz are obtained with the following rotor speeds and number of poles (using Formula A66 above):

TABLE A 53 :
Number of poles and synchronous speeds of three-phase motors and generators

Number of poles	50 Hz grid frequency	60 Hz grid frequency
2	3000 rpm	3600 rpm
4	1500 rpm	1800 rpm
6	1000 rpm	1200 rpm
8	750 rpm	900 rpm
10	600 rpm	720 rpm

Generally, generators with as few poles as possible are to be chosen in order to reduce size and costs of the machine. However, the speed is usually determined by the prime mover: Waterwheels for example require a large number of poles, because of their low speed while high speed applications such as Pelton turbines under high heads may make do with 2 or 4 pole generators even if direct-coupled.
On the other hand, low speed generators with many poles may not be affected in their structural integrity when the turbine/generator unit accelerates to runaway speed. Runaway speeds of turbines are in the order of 1.5 to 2 times nominal speed. A two-pole generator may therefore accelerate up to 6000 rpm at full load rejection and bearings and rotating elements are likely to be damaged at such speeds; whereas low speed generators with many poles may even rotate at runaway speed for extended periods without being damaged.

8.5. Three-Phase Systems and Consumer Load

Micro-hydropower of a certain size (beyond 10 to 15 kW) usually use three-phase four wire distribution systems with the generator connected in Y (advantages see section 8.1 above). Such a system provides 6 options of connecting single-phase consumers and 2 options for three-phase consumers. Figure A54 gives an overview of the possible connections in a three-phase system including a neutral wire connected to the star point of the generator.
Connecting a generator in delta is only used for small systems supplying exclusively three-phase consumers at short distances from the generator.

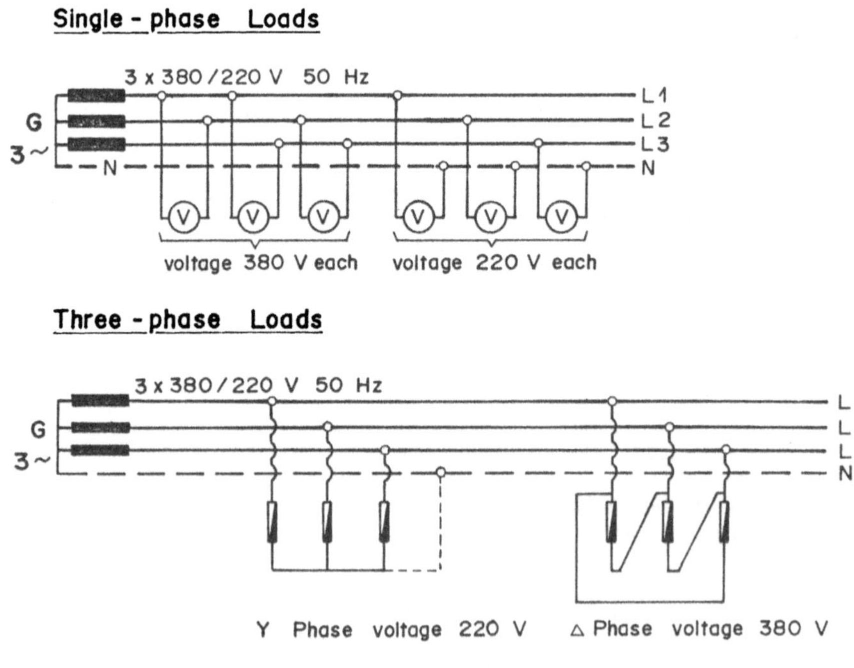

FIGURE A 54 :
Overview of possible connections of consumers and corresponding voltages in a three-phase four-wire system

In a three-phase system, several single-phase appliances together (lighting, other small household appliances) constitute a three-phase consumer. It is important that these single-phase consumers are connected to the grid in such a way that all three phases are equally loaded. The three-phase generator is designed to deliver a third of the total power through each phase. Imbalanced loading of phases (e.g. all lamps of a whole village on phase 1) may create a situation where excess power flows through one or two phases, overheating the windings and finally causing the insulation to burn and the generator to fail.
In an imbalanced system with the neutral connected to the starpoint , a fairly high current may flow back to the power station. Eliminating imbalances is very difficult for small grids because relatively large single-phase appliances connected to the same phase may be switched on simultaneously overloading the generator (that is why small grids preferably use single-phase systems). In larger grids, imbalances on phases may equal each other out according to the laws of probability provided that the single-phase appliances are well distributed over the three phases.
The current in the neutral indicates whether the system is well balanced or not:

There is no current in the neutral wire in a balanced three-phase system

The current in the neutral wire can be calculated by adding graphically or analytically the vectors of the three line currents of the system (see example 5 below).

Example 5:

A three-phase four-wire system (380/220 V ; 50 Hz) has the following loads:

- three-phase motor connected in Y with the following name plate data:
  U = 380 / 220 V (Y / Δ)
  I = 30 / 52 A
  P_N = 15 kW
  cos ϕ = 0.85

- resistive load (electrical stove) between L1 and Neutral: 2.5 kW
  between L2 and Neutral: 5.5 kW

- capacitive load and resistive load in series between L3 and Neutral: R = 5 Ω and C = 318 μF.

Calculate the phase currents of the appliances, the line currents and the current in the neutral wire.

8.6 Power Factor Correction in Three-Phase Systems

Basically, power factor correction in three-phase systems is similar to the procedure shown for single-phase networks; the same formulae as given in section 7.11 above apply. The only difference is that in a three phase system the capacitors may be connected in Y or in delta. From formula A65, we can see that the capacitance C_Δ is always smaller than C_Y (for capacitor $Z_C = X_C = 1/\omega C$). That is why capacitors for power factor correction are usually connected in delta despite the higher voltage and thus higher insulation class required.
Using the following general indications, the required capacitors can immediately be determined from the reactive power to be compensated per phase (valid for a 380/ 220 V; 50 Hz network)

Compensating reactive power of **1 kvar in Y - connection** (220 V / 50 Hz) requires capacitors of **66 μF**
Compensating reactive power of **1 kvar in Δ - connection** (380 / 50 Hz) requires capacitors of **22 μF**

Example 6: A three-phase motor is connected in Y to a three-phase four-wire system (380/220 V; 50 Hz). The name plate provides the following data:

$$U = 380\,/\,220\text{ V} \quad (Y\,/\,\Delta)$$
$$I = 30\,/\,52\text{ A}$$
$$P_N = 15\text{ kW}$$
$$\cos\phi = 0.85$$

The power factor should be corrected to $\cos\phi = 0.915$. Calculate the reactive power to be provided by the capacitors and the size of the capacitors in delta connection.

9. VOLTAGE DROP AND CONDUCTOR SIZE

9.1 General

For short supply lines, the conductor size is based on keeping heating of the conductors within acceptable limits (see Table in Appendix B1). For longer lines, it is the maximum acceptable voltage drop which determines the conductor size. Some loads / consumers are sensitive to large voltage variations (e.g. motors) while others are less (e.g. incandescent lamps). If no details of the future appliances are available, the following **maximum voltage drops** may be used to determine the conductor size:

- **House and factory installations:**
  - For main lines as far as the main fuse in buildings: max 0.5 %

  - For light installations: max 1.5 %

  - For power installations: max 3 %

- **Distribution lines:**
  - 10 % (if the admissible voltage variation of appliances is ± 10 %)

9.2 Voltage Drop in DC

Referring to Ohm's law U = R I and the formula for resistances R = ρ l / A (see section 2.1 above), we can immediately calculate the voltage drop of a line in DC:

$$U_V = \frac{2\rho\, l\, I}{A} \qquad \text{(A67)}$$

Where U_v = voltage drop in the conductor in V (< permissible voltage drop as indicated under 9.1 above)

ρ = resistivity of the conductor material (Ω mm2/m)
l = **single** length of the conductor between source and end-user (m)
I = current (A)
A = conductor cross sectional area (mm2)

Note that the factor 2 allows for the voltage drop on the line from the appliance back to the voltage source.

Or solving the equation for the minimum cross sectional area required to keep the voltage drop U_V below the permissible value:

$$A = \frac{2\rho\, l\, I}{U_v} \qquad \text{(A68)}$$

9.3 Voltage Drop in Single-Phase AC

9.3.1 Considering line resistance only

Note that the following formulae are only correct if the inductance of the cables and conductors is negligibly small as compared to their resistance. This applies to most low-voltage (LV) house wiring and factory installations but not necessarily to distribution lines in micro-hydropower systems. For the latter cases, we recommend that the cables should be pre-selected using the formulae for ohmic losses but the final selection should be made only after the total voltage drop including the reactance of the cables has been checked (using Formula A71).

An additional problem arises for inductive loads: the voltage drop of the cable (line) is not in phase with the voltage of the consumer. *Figure A55 below shows the circuit diagram of a single-phase motor (inductive* consumer) connected to an AC generator. Due to the voltage drop on the line, the motor does not receive full generator voltage. Unlike the DC case, only that part of the total voltage drop in the conductor (U_{wire}) which is in phase with the consumer voltage must be considered as the effective voltage drop (U_v) for the motor. This is illustrated with the phasor diagram in Figure 55 below. Since U_{wire} is usually small as compared to the voltage drop across the consumer, we can calculate U_v approximately as follows (assuming that AB $\approx$ AC on Figure A 55):

$U_V = U_{wire} \cos\phi$. Since U_{wire} is equal to U_V in DC (see Formula 67 and 68) we can write:

$$\boxed{A = \frac{2\ \rho\, l\ \ I\ \cos\phi}{U_v}} \qquad \text{(A69)}$$

where U_v = permissible voltage drop for the application

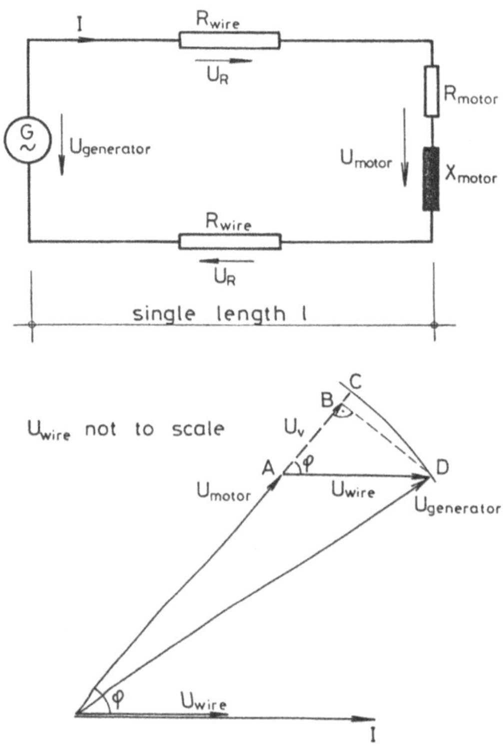

FIGURE A 55 :
Voltage drop in a single-phase AC system with an inductive load

9.3.2 Considering resistance and inductance of a line

If the reactance of a cable becomes important, the following formula may be applied:

$$\boxed{L = 0.05 + 0.2\ \ln(\frac{d}{r})} \qquad \text{mH/km (milli-Henry / km)} \qquad \text{(A71)}$$

where d = distance between two conductors and
r = radius of a conductor

and $X = L * 2\,\pi\, f\ [\Omega / \text{km}]$

Refer to manufacturers brochures for the thickness of insulation materials between phases in composite cables.

For a 50 Hz grid, the reactance X of overhead lines is about 0.4 Ω / km and for cables 0.07 to 0.17 Ω / km

The total voltage drop can be found using the two components $U_{R,wire} = 2 * R * I$ (in phase with I) and $U_{X,wire} = 2 * X * I$ (90° out of phase with I) and adding them graphically (see also Figure A56 below). Note that the factor 2 allows for the voltage drop on the line from the appliance back to the voltage source.
Assuming that $\phi_{load} \cong \phi_{total} = \phi$, we can write:

Total voltage drop $U_v = 2\,(I\,R\cos\phi + I\,X\sin\phi)$ < permissible voltage drop for the application

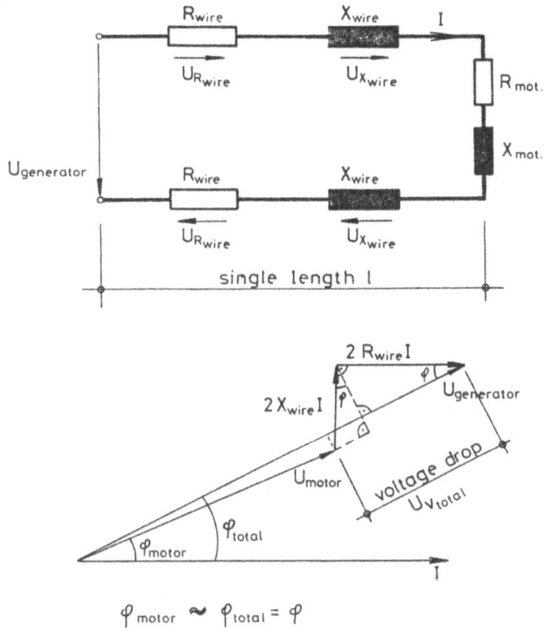

FIGURE A56 :
Total voltage drop in a single-phase system including resistance and reactance of the supply line

9.4 Voltage Drop in Three-Phase AC

Three-phase systems are basically 3 single-phase systems with only one return wire. It is usual practice to assume balanced loads for the calculation of the voltage drop, i.e. the neutral wire is without current. Therefore, the same formulae as for single-phase systems apply but calculated with the single length of the line:

- considering line resistance only:

$$A = \frac{\rho\, l\, I \cos\phi}{U_v} \qquad \text{(A70)}$$

- considering line resistance and reactance

$$U_v = R\,I\cos\phi + X\,I\sin\phi \qquad \text{(A71)}$$

Example 7:
In an irrigation project, a pump with a 30 kW three-phase motor has to be connected to a three-phase network (380/220 V , 50 Hz). The pump house is 250 m away from the switchboard. The motor has a cosϕ of 0.87 and an efficiency $\eta = 89$ %. Admissible voltage drop is 3 %. What size of alu cable is to be chosen? ($\rho_{alu} = 0.03\ \Omega\ mm^2/m$ at 20° C)

10. ANSWERS TO THE EXAMPLES

10.1 Example 1 / page 120

Resistances for each appliance / device:

- Lamps:

$$R_1 = \frac{U_n^{\ 2}}{P_n} = \frac{24^2}{25} = 23\ \Omega$$

$$R_2 = \frac{24^2}{10} = 57.6\ \Omega$$

$$R_3 = \frac{24^2}{15} = 38.4\ \Omega$$

$$R_4 = R_5 = \frac{12^2}{15} = 9.6\ \Omega$$

- Wiring:

$$\rho_{Cu20^\circ C} = 0.0175\ \Omega\ mm^2/m$$

$$R_{Cu20^\circ C} = \frac{\rho\ l}{A} = \frac{0.0175 * 2 * 200}{1.5} = 4.67\ \Omega$$

$$R_{Cu35^\circ C} = R_{Cu20}(1 + \frac{35 - 20}{235 - 20}) = 5.0\ \Omega$$

- Circuit diagram and equivalent circuit

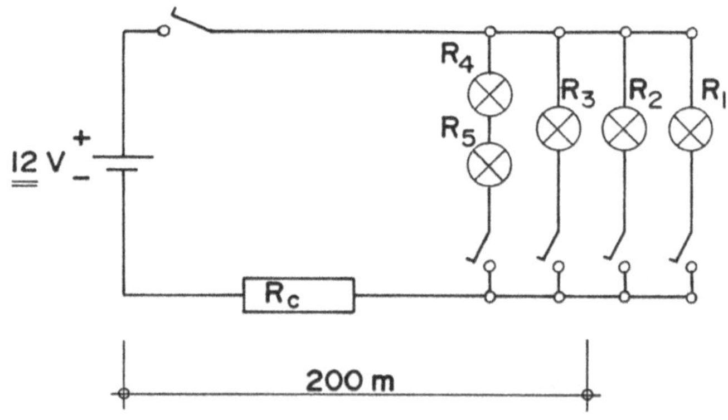

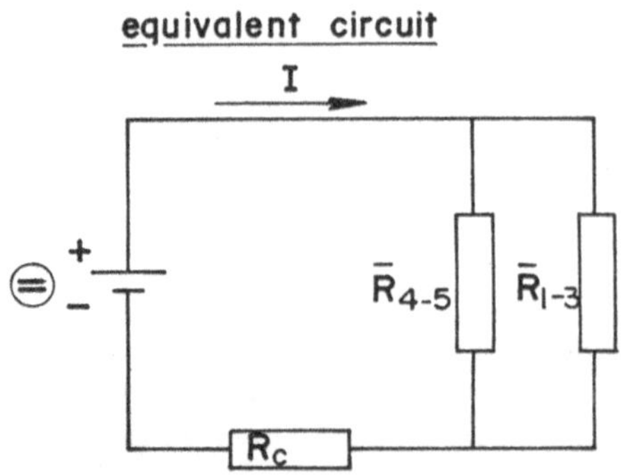

- Equivalent resistances:

series: $R_{4\text{-}5} = R_4 + R_5 = 2 * 9.6\ \Omega = 19.2\ \Omega$

parallel: $R_{1\text{-}3} = \frac{1}{1/R_1 + 1/R_2 + 1/R_3} = \frac{1}{1/23 + 1/57.6 + 1/38.4} = 11.5\ \Omega$

parallel: $R_{1\text{-}5} = \frac{1}{1/R_{1\text{-}3} + 1/R_{4\text{-}5}} = \frac{1}{1/11.5 + 1/19.2} = 7.2\ \Omega$

series: $R = R_{1\text{-}5} + R_{Cu} = 7.2 + 5.0 = 12.2\ \Omega$

- Voltage available: U = 12 V;
- Current in the supply line: $I = U/R = 12 / 12.2 = 1$ A
- Power requirement: $P = U * I = 12 * 1 = 12$ W

- Conductor size: 1. Keeping heat within acceptable limits (see Table B1, main volume)
Maximum current for a 1.5 mm^2 section: 10 A > I_{actual} = 1 A

2. Voltage drop
$U_v = R_c * I = 5.0 * 1.0 = 5$ V ==> 42 % of U_{source}
far too much ==> use a larger cross section (see also Chapter 9).

- Available voltage on the lamps:

$U_{lamps\ 1\text{-}3} = R_{1\text{-}5} * I = 7.2 * 1.0 = 7.2$ V
(very low as compared to their rating of 24 V ==> will not light up)
$U_{lamps\ 4\text{-}5} = 7.2 / 2 = 3.6$ V
(also too low for their rating of 12 V)

- Conclusion: The example shows that a 12 Volt DC source is insufficient to supply a house with a number of loads at 200 m distance. A larger conductor size may improve the situation but will most probably pose economical problems.

10.2 Example 2 / page 124

$$C = \frac{2\ W}{U^2} = \frac{2 * 100}{800^2} = 3.13 * 10^{-4} = 313\ \mu F$$

circuit diagram:

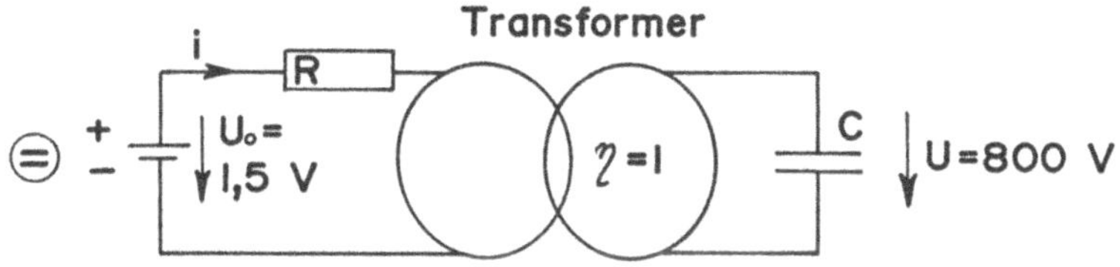

Maximum current at the beginning of the charging process:

$$i_o = i_{max} = \frac{U_o}{R} < 0.1\ A \implies R_{req} > \frac{U_o}{i_{max}} = \frac{1.5}{0.1} > 15\ \Omega$$

10.3 Example 3 / page 125

- The second capacitor is twice as big as the first one: $2\,C_1 = C_2$
- For parallel connection: $C_{total} = C_1 + C_2 = C_1 + 2C_1 = 3\,C_1$

$$C_1 = C_{total}/3 = 104\ \mu F \quad ; \quad C_2 = 2\,C_1 = 208\ \mu F$$

- For series connection:

$$C_{total} = C_{equi.} = \frac{1}{1/C_1 + 1/C_2} = \frac{1}{1/C_1 + 1/2C_1} = 2C_1/3$$

$C_1 = 3\,C_{total}/2 = 470\ \mu F$ $\qquad C_2 = 2\,C_1 = 940\ \mu F$

The parallel connection is to be preferred.

10.4 Example 4 / page 164

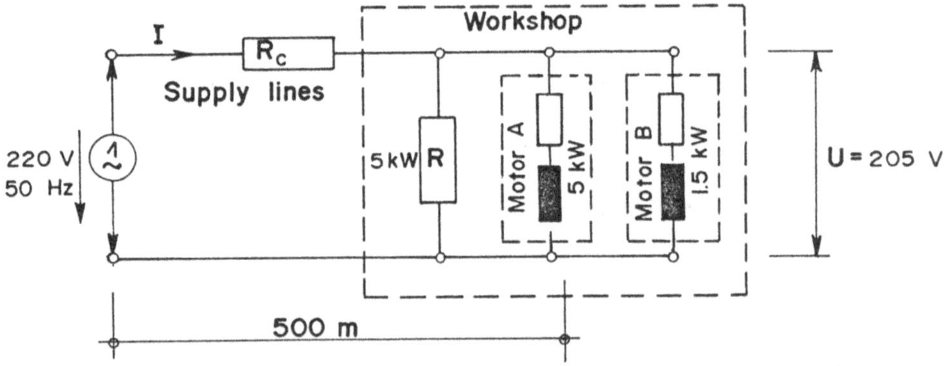

- Individual power factor correction for the two motors:

Motor A: $P_A = \frac{P_n}{\eta} = \frac{5.0}{0.8} = 6.25 \text{ kW}$

$$C_A = \frac{P_A (\tan\phi_1 - \tan\phi_2)}{U^2 2\pi f} = \frac{6250 \{\tan[\arccos(0.75)] - \tan[\arccos(0.9)]\}}{205^2 2\pi 50} = 188 * 10^{-6} F = 188 \mu F)$$

Note, in practice, U is usually taken as nominal voltage (220V) to calculate the capacitance C

Motor B: $P_B = \frac{P_n}{\eta} = \frac{1.5}{0.75} = 2.0 \text{ kW}$

$$C_B = \frac{P_B (\tan\phi_1 - \tan\phi_2)}{U^2 2\pi f} = \frac{2000 \{\tan[\arccos(0.7)] - \tan[\arccos(0.9)]\}}{205^2 2\pi 50} = 81 * 10^{-6} F = 81 \mu F)$$

- Total power of the workshop:

- lamps, electric stove (pure resistances): $P_1 = S_1 = 5$ kW; $Q_1 = 0$

- motor A: $P_A = 6.25$ kW $S_A = P_A / \cos\phi = 6.25 / 0.9 = 6.94$ kVA (after power factor correction)
$Q_A = S_A \sin\phi = 6.94 \sin[\arccos(0.9)] = 3.0$ kVar

- motor B: $P_B = 2.0$ kW $S_B = 2.0 / 0.9 = 2.22$ kVA (after power factor correction)
$Q_A = S_A \sin\phi = 2.22 \sin[\arccos(0.9)] = 0.97$ kVar

- Power triangle (vector addition)

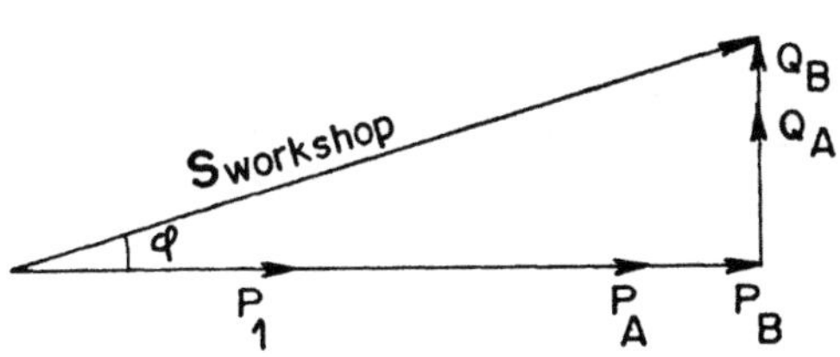

$\Sigma P = 13.25$ kW (without power loss in supply line)
$Q = 3.97$ kVar
$S_{workshop} = \sqrt{13.25^2 + 3.97^2} = 13.8$ kVA
$\cos\phi_{workshop} = \Sigma P / S = 13.25 / 13.8 = 0.96$
$I' = S_{workshop} / U'_{workshop} = 13'800 / (220 - 15) = 67.3$ A (preliminary value; actual voltage drop depends on the cable selected for the supply lines)

- Conductor size required for the supply lines (see also Chapter 9)

admissible voltage drop at workshop: $U_v = 15$ V (in phase with workshop voltage)

==> voltage drop in supply lines: $U_{wire} < U_v / \cos\phi_{workshop} = 15 / 0.96 = 15.6$ V

$$R_c < U_{wire} / I = 15.6 / 67.3 = 0.232\ \Omega$$

resistivity of copper : $\rho_{20^\circ C} = 0.0175\ \Omega mm^2/m$

$$A > \frac{\rho\, l}{R_c} = \frac{0.0175 * 2 * 500}{0.232} = 75.5\ mm^2$$

From Table B1: diameter = 12.5 mm, A = 95 mm^2

(Note that in this case, the criterion for the selection of the cable is not heating but voltage drop.)

$$R_{actual} = \frac{\rho\, l}{A} = \frac{0.0175 * 2 * 500}{95} = 0.184\ \Omega$$

Total voltage drop $U'_{wire} = R_{actual} * I' = 0.184 * 67.3 = 12.4$ V (in phase with I' but not with $U'_{workshop}$)

Phasor diagram:

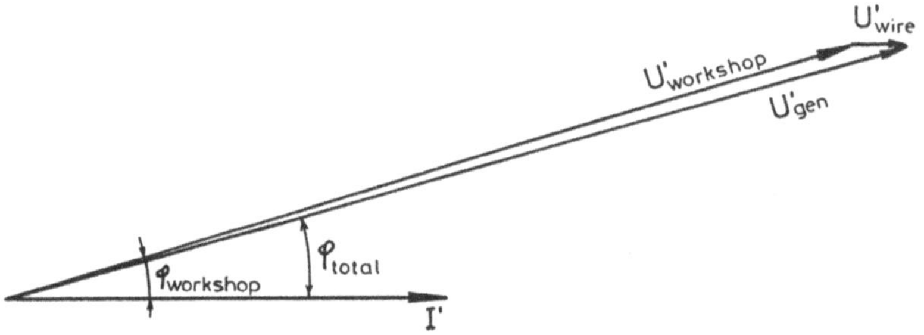

$\phi_{workshop} = \arccos(0.96) = 16.3^\circ$

$$U'_{gen} = \sqrt{(U'_{workshop} \sin\phi_{workshop})^2 + (U'_{workshop} \cos\phi_{workshop} + U'_{wire})^2} =$$
$$U'_{gen} = \sqrt{(205 * \sin 16.3^\circ)^2 + (205 * \cos 16.3^\circ + 12.4)^2} = 217\ V$$

Voltage available at the generator is 220 V

$I_{total} \cong I' * U/U' = 67.3 * 220/217 = 68.2$ A

Total apparent power $S = U * I_{total} = 220 * 68.2 = 15$ kVA (at generator)

Power factor at generator (all appliances connected):
(from phasor diagram above)

$$\tan\phi_{total} = \frac{205 * \sin 16.3^\circ}{205 * 0.96 + 12.4} = 0.275$$

==> $\phi_{total} = \arctan(0.275) = 15.4^\circ$ and $\cos\phi_{total} = 0.96$

10.5 Example 5 / page 173

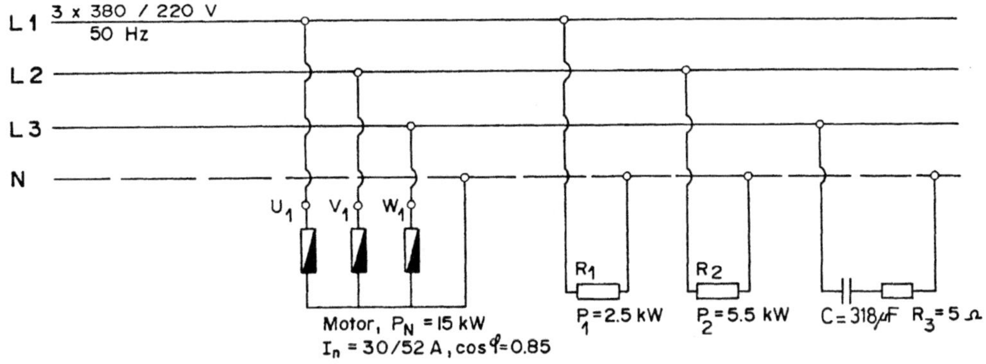

1. Determine the currents for each phase as for a single-phase system

<u>**Equivalent circuit**</u>

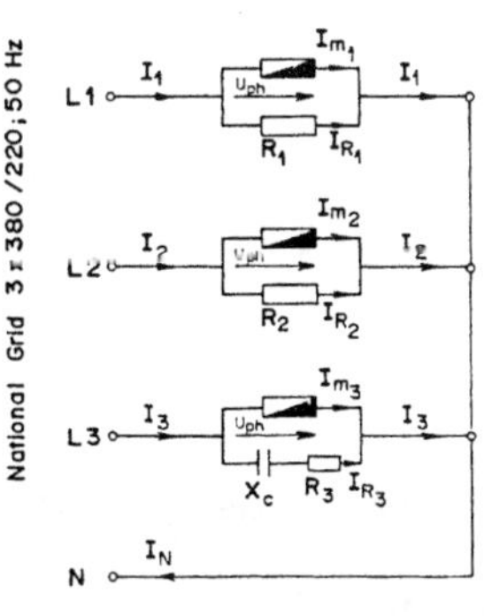

- Motor in Y (from name plate):

  $I_{m_1} = I_{m_2} = I_{m_3} = 30$ A; $\cos\phi_m = 0.85$ (ind.)

  $\sin\phi_m = +0.527$ (ind.) ; $\phi_m = 31.79°$ (ind.)

- Stove ($\cos\phi = 1$)

  $I_{R_1} = P/U_{ph} = 2500 / 220 = 11.4$ A ; $\phi = 0$

  $I_{R_2} = P/U_{ph} = 5500 / 220 = 25$ A ; $\phi = 0$

- Resistance and capacitance in phase 3:

  $X_C = 1 / (\omega C) = 1 / (2 \pi f C) = 10\ \Omega$

from Table 40: $Z = \sqrt{R_3^{\ 2} + X_C^{\ 2}} = \sqrt{5^2 + 10^2} = 11.18\ \Omega$

$I_{R_3} = I_{X_C} = U_{ph} / Z = 220 / 11.18 = 19.68$ A

$\tan\phi = X_C / R_3 = 10 / 5 = 2.0$; $\phi_C = 63.43°$ (cap.)

$\cos\phi_C = 0.447$ (cap.) ; $\sin\phi_C = -0.894$ (cap.)

2. Resultant currents per phase (to be determined with phasor diagram)

- Phase 1

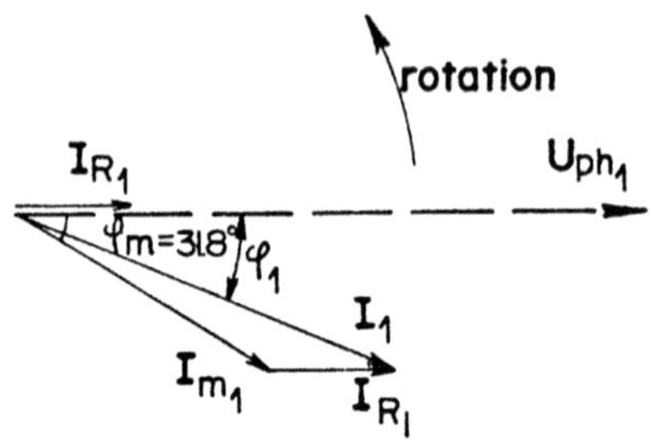

$$I_1 = \sqrt{(I_{m_1}\cos\phi_m + I_{R_1})^2 + (I_{m_1}\sin\phi_m)^2} = 40.1\ \text{A}$$

$$\cos\phi_1 = \frac{I_{m_1}\cos\phi_m + I_{R_1}}{I_1} = 0.920\ \text{(ind.)} \qquad \phi_1 = 23.0°$$

- Phase 2

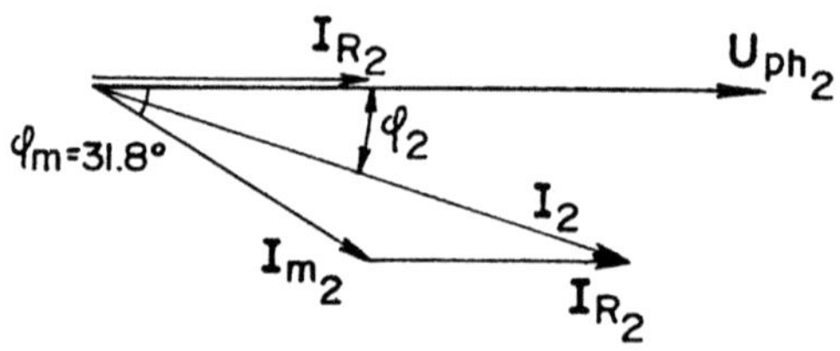

$I_2 = 52.9$ A $\qquad \cos\phi_2 = 0.954$ (ind.) $\phi_2 = 17.4°$ (ind.)

- Phase 3:

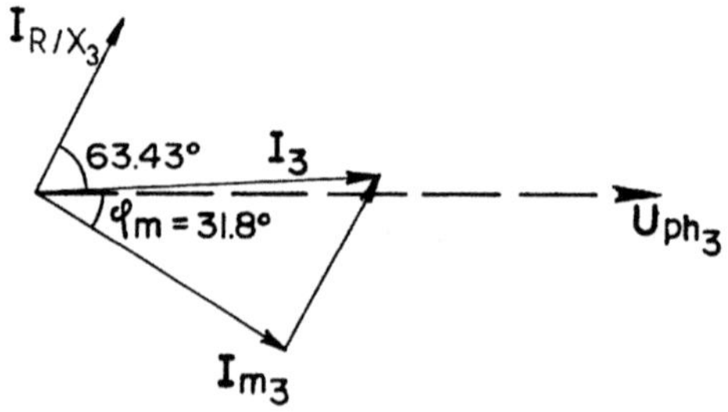

$$I_3 = \sqrt{(I_{m_3}\cos\phi_m + I_{R/X_3}\cos\phi_C)^2 + (I_{m_3}\sin\phi_m - I_{R/X_3}\sin\phi_C)^2} = 34.3\ \text{A}$$

$\phi_3 = 3.0°$ $\qquad \cos\phi_3 \cong 1$

3. Current in neutral

- Phasor diagram for voltage: $U_{ph} = 220$ V, each phase 120° displaced

- Line (= phase) currents to be plotted into the voltage phasor diagram taking into account the different phase angles ϕ (leading or lagging):

inductive => lagging voltage
capacitive => leading voltage

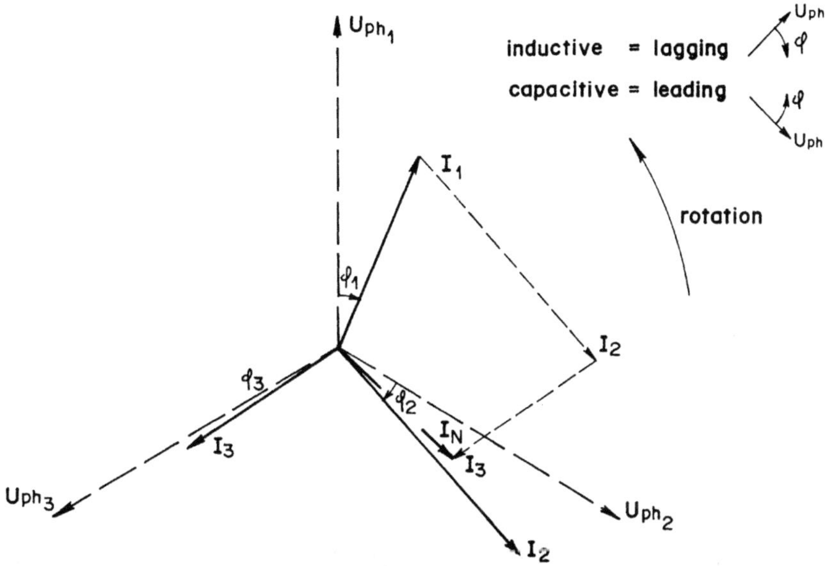

$I_{neutral} \cong 31$ A (relatively high imbalance)

- Refer to max. admissible imbalances of generators indicated by manufacturers.

4. Total power of all consumers together

- active power for non-symmetrically loaded systems: $P_{total} = P_{ph_1} + P_{ph_2} + P_{ph_3}$

$P_{ph_1} = U_{ph}\ I_{ph_1} \cos\phi_1 = 220 * 40.1 * 0.920 =$ 8109 W

$P_{ph_2} = U_{ph}\ I_{ph_2} \cos\phi_2 = 220 * 52.9 * 0.954 =$ 11'105 W

$P_{ph_3} = U_{ph}\ I_{ph_3} \cos\phi_3 = 220 * 34.3 * 1.0 =$ 7536 W

$P_{total} =$ 26'750 W

- similarly for total reactive power: $Q_{total} = Q_{ph_1} + Q_{ph_2} + Q_{ph_3}$

with $Q_{ph} = U_{ph}\ I_{ph} \sin\phi$

$Q_{total} = 7.35$ kVar

- total apparent power

$$S_{total} = \sqrt{P_{total}^2 + Q_{total}^2} = 27.74 \text{ kVA}$$

10.6 Example 6 / page 174

The efficiency of the motor is not known, i.e. the electrical power must be determined using voltage and current indications:

$P_{el} = \sqrt{3}\, U_{line}\, I_{line} \cos\phi = \sqrt{3} * 380 * 30 * 0.85 = 16'800$ W

- Capacitors in Δ : (each capacitor will receive line voltage $U_{line} = 380$ V)

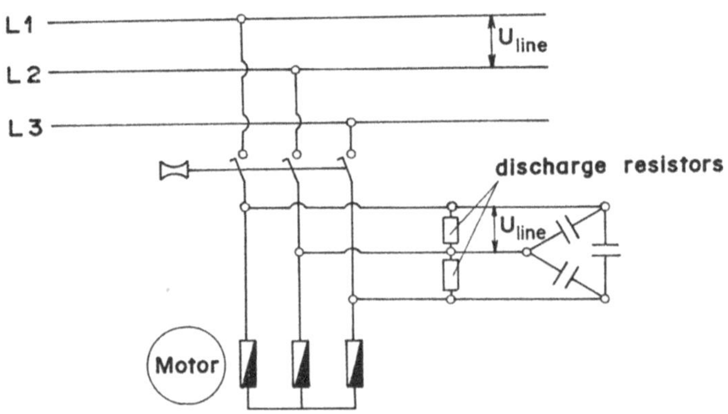

$$C_{\Delta/phase} = \frac{P(\tan\phi_1 - \tan\phi_2)}{3\, U^2\, 2\pi f} = \frac{16'800\ \{\tan[\arccos(0.85)] - \tan[\arccos(0.915)]\}}{3 * 380^2\, 2\pi\, 50} = 22 * 10^{-6}\,F = 22\ \mu F$$

10.7 Example 7 / page 177

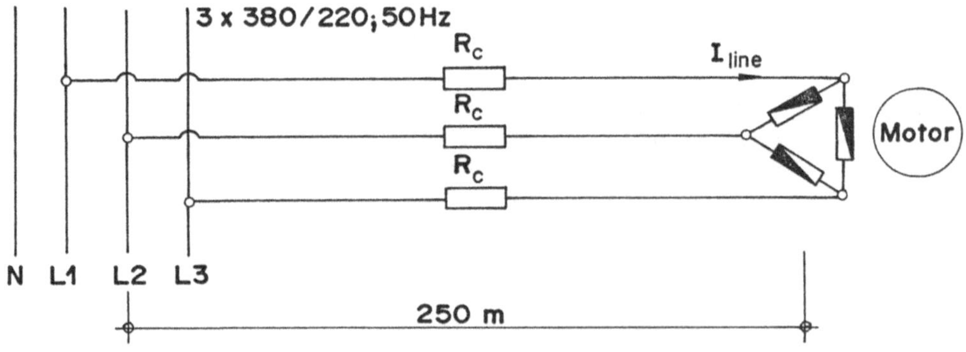

$P_{el} = P_n / \eta = 30 / 0.89 = 33.71$ kW

$P_{el} = \sqrt{3}\, U_{line}\, I_{line} \cos\phi_m \implies I_{line} = \dfrac{P_{el}}{\sqrt{3}\, U_{line} \cos\phi_m} = \dfrac{33'710}{\sqrt{3} * 380 * 0.87} = 58.87$ A

- admissible voltage drop U_v = 3% of $U_n = 0.03 * 380 = 11.4$ V

- voltage drop in phase with current I : $U_{wire} = U_v / \cos\phi_m = 11.4 / 0.87 = 13.1$ V

$R_{alu} < U_{wire} / I_{line} = 13.1 / 58.87 = 0.223\ \Omega$

$A > \rho_{alu} * l / R_{alu} = 0.03 * 250 / 0.223 = 33.7\ mm^2$

- Cable selection (see Appendix B1):
  criterion heating: I = 58,9 A ==> $A > 16\ mm^2$
  criterion voltage drop: $A > 33.7\ mm^2$

- Pre-selection: cable diameter : 9 mm $A = 50\ mm^2$ per phase

- Verification of total voltage drop including the reactance of the cable:
  from manufacturer's brochure:

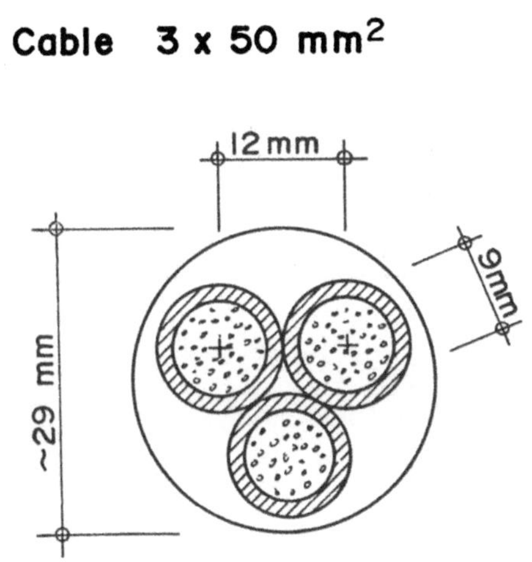

$L = 0.05 + 0.2 \ln(12 / 4.5) = 0.246$ mH / km

$X_{wire} = \omega L = 2\pi\, 50 * 0.246 * 10^{-3} = 0.077\ \Omega$ / km or for l = 250 m ==> $X = 0.02\ \Omega$

$R_{wire} = \rho_{alu} * l / A = 0.03 * 250 / 50 = 0.15\ \Omega$

Total voltage drop $U_v \cong I*R*\cos\phi + I*X*\sin\phi =$

$58.9 * 0.15 * 0.87 + 58.9 * 0.02 * \sin[(\arccos(0.87)] =$

$7.69 + 0.56 = 8.25$ V $< U_v$ adm. = 11.4 V

APPENDIX B

SELECTED DIAGRAMS AND TABLES FOR ELECTRICAL ENGINEERING DESIGN

Appendix B1

Standard Wire and Cable Sections and Permissible Currents (Heating Limit)

Standard Wire Section (mm2)	Wire Diameter (mm)	Cable Diameter (mm)	Maximum Current Stationary Installation (A)	Maximum Current Moving Installation (A)
0.75	1.0			6
1	1.13		6	10
1.5	1.4		10	15
2.5	1.8		15	20
4	2.25		20	25
6	2.8		25	40
10	3.6	4	40	60
16	4.5	5.1	60	80
25		6.3	80	100
35		7.5	100	
50		9	125	
70		11	150	
95		12.5	200	

Appendix B2:

End-Use Appliances and their Approximate Power Demand

Appliances	Nominal power in Watt
Lighting:	
- Incandescent lamps	25 - 200
- Fluorescent tubes	25 - 100
- Energy-efficient lamps	8 - 25
Heating:	
- Boiler 100 l	2400
- Boiler 1000 l	8000
Household appliances:	
- Cooker / stove (single hotplate)	1000 - 2000
- Cooker / stove (3 hotplates)	6000 - 7000
- Low-wattage cooker	200
- Refrigerator 180 litre	100
- Deep-freezer 300 litre	200
- Washing machine	2100 - 3300
Electronics:	
- Radio (portable)	10 - 100
- Radio (stationary)	100 - 300
- TV (colour)	200
- Computer	200 - 250
- Printer	80 - 150
Workshop:	
- Circular saw	2500 - 5500
- Centre lathe	1500 - 5000
- Universal milling machine	1500 - 3000
- Bench and pillar drilling machine	500 - 1500
- Angle grinder	500 - 1500

Source:
- FAT, Switzerland: Rapport 348
- International Labour Office, Geneva: Equipment planning guide

Appendix B3

Efficiencies and Power Factors for Induction Motors

(average values of standard motors)

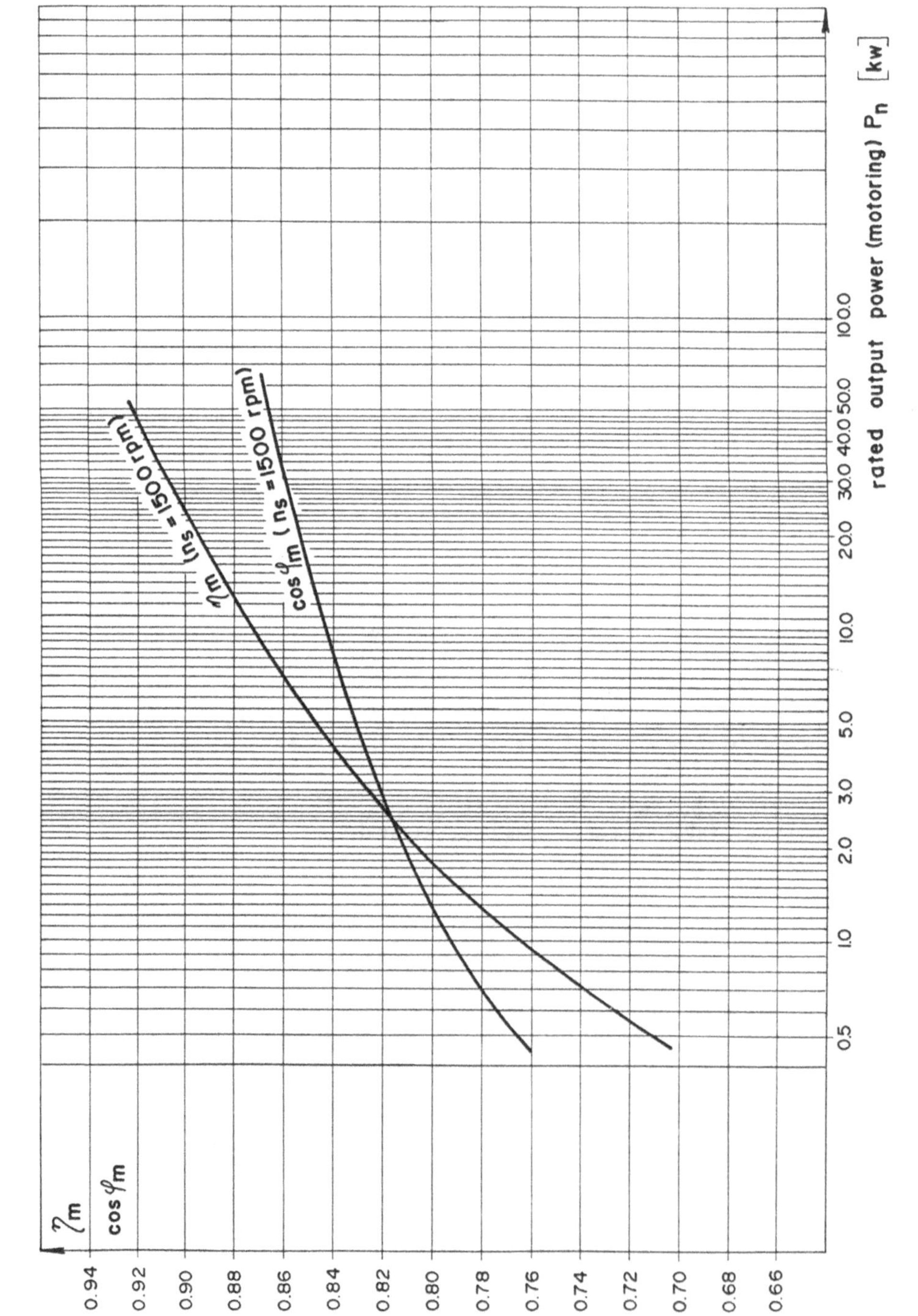

Appendix B4

Slip versus Mechanical Power for Induction Machines
(for "hard" non-saturated motors only)

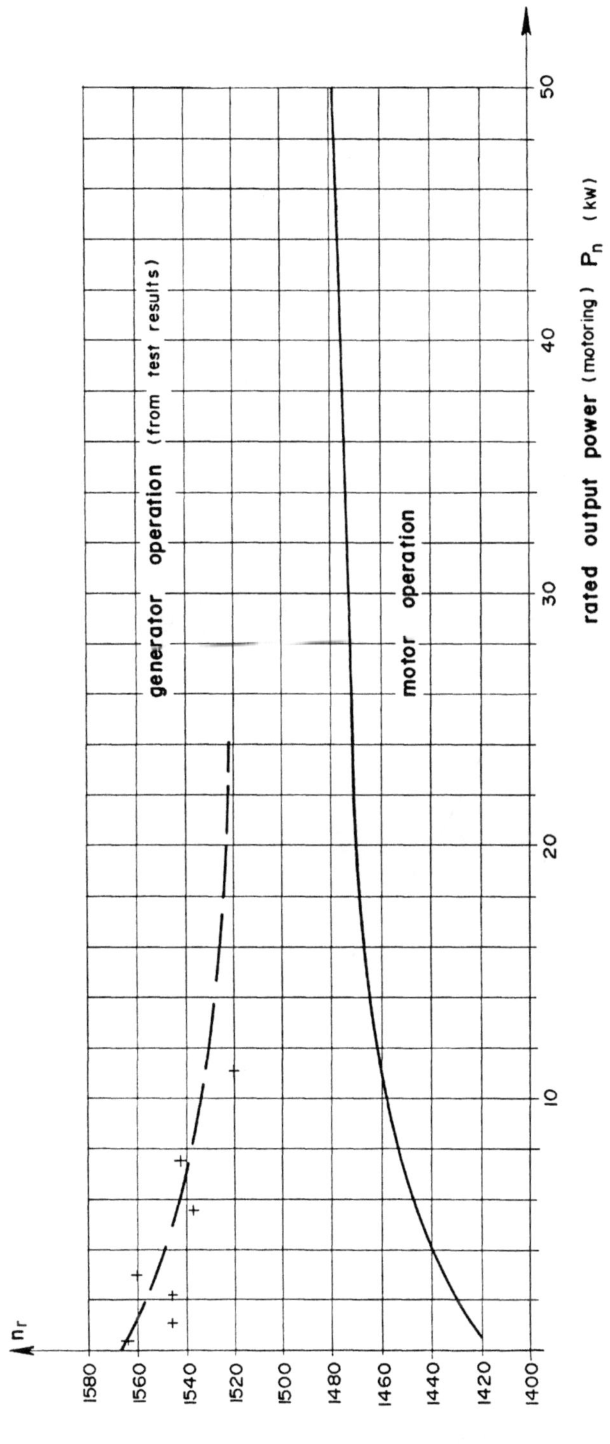

Appendix B5

General Range of Efficiencies for Synchronous Generators

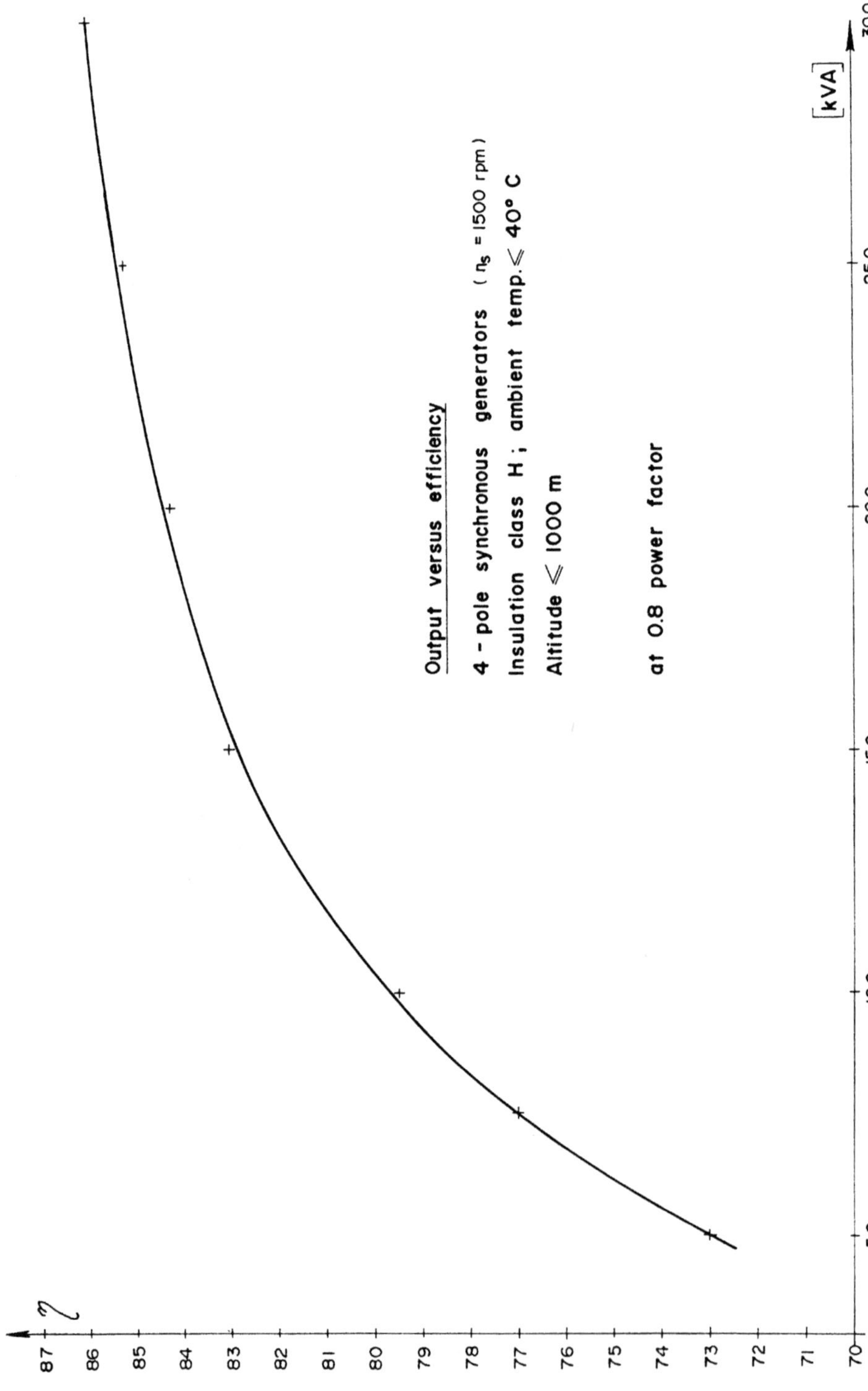

Appendix B6

General Range of Efficiencies for Couplings

Type of coupling		*maximum gearing ratio (single stage)*	*max. power per stage in kW*	*general range of efficiency (per stage)*	*advantage / disadvantage*
Flat belt		5	> 100	96 - 98 %	+ simple - high initial tensioning (forces on shafts) - careful alignment required
V-belts		8	35	94 - 97 %	+ readily available + quiet + do not require very precise alignment - high initial tensioning
Toothed belts (timing belts)	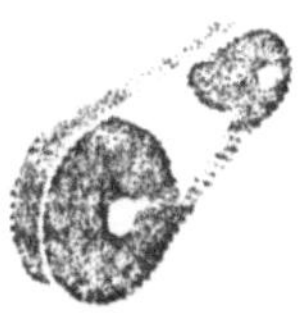			98 %	+ high efficiency + no slip + little initial tensioning - noisy - expensive - not readily available
Chain drives		6	200	97 - 98 %	+ high efficiency + no slip + no initial tensioning - expensive - maintenance (lubrication)
Gear box: Spur gears		8		96 - 98 %	+ require less space than belts - expensive - maintenance
Gear box: Bevel gears		6		96 - 98 %	+ connect intersecting shafts - expensive - maintenance

Source:
- *Niemann G.* : Maschinenelemente, Vol. II, 1965
- *Dubbel:* Taschenbuch für den Maschinenbau, 1986
- various manufacturers' brochures

Appendix B7

Turbine Types and Specific Speeds (in function of head and flow)

(adapted from *Bundesamt für Wasserwirtschaft:* Kleinwasserkraftwerke in der Schweiz, Part I, 1983)

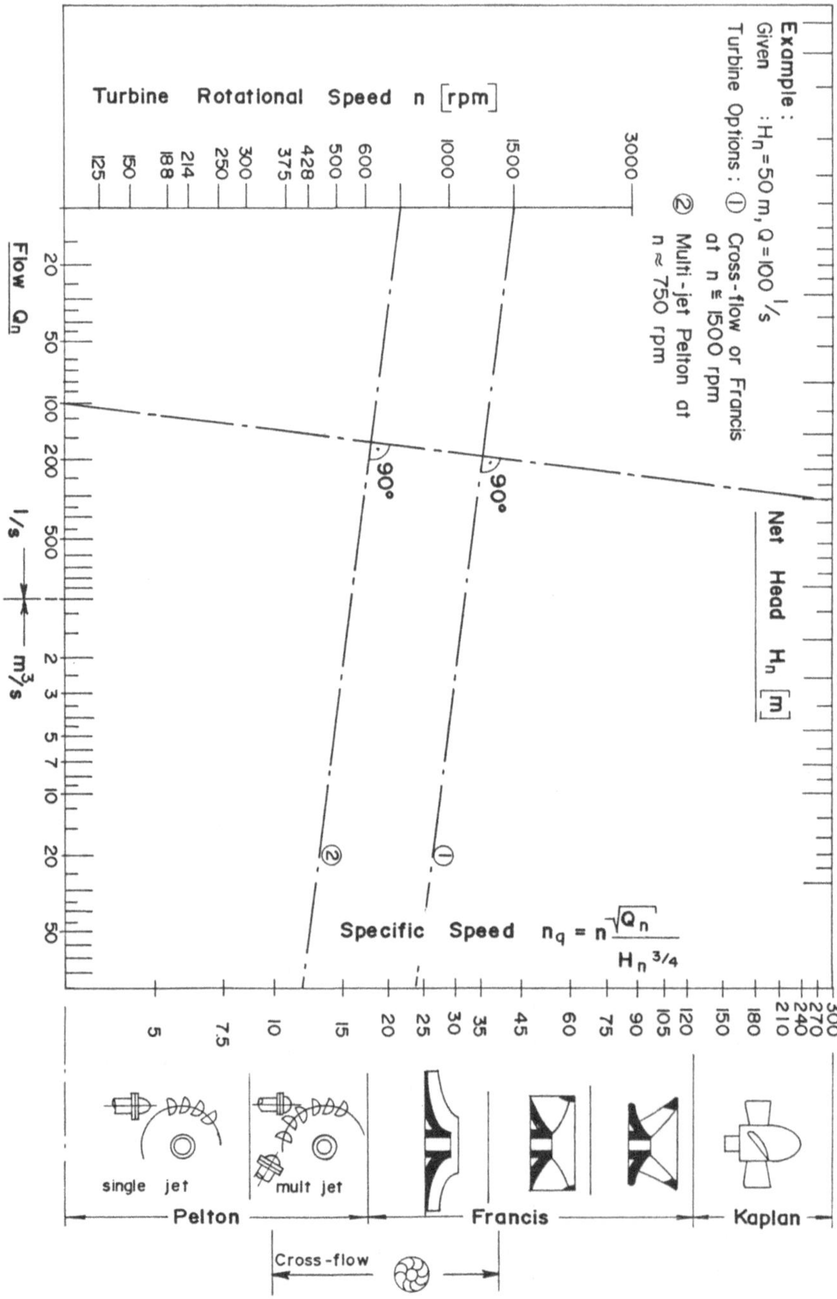

Appendix B8

Turbine Efficiencies at Nominal Flow

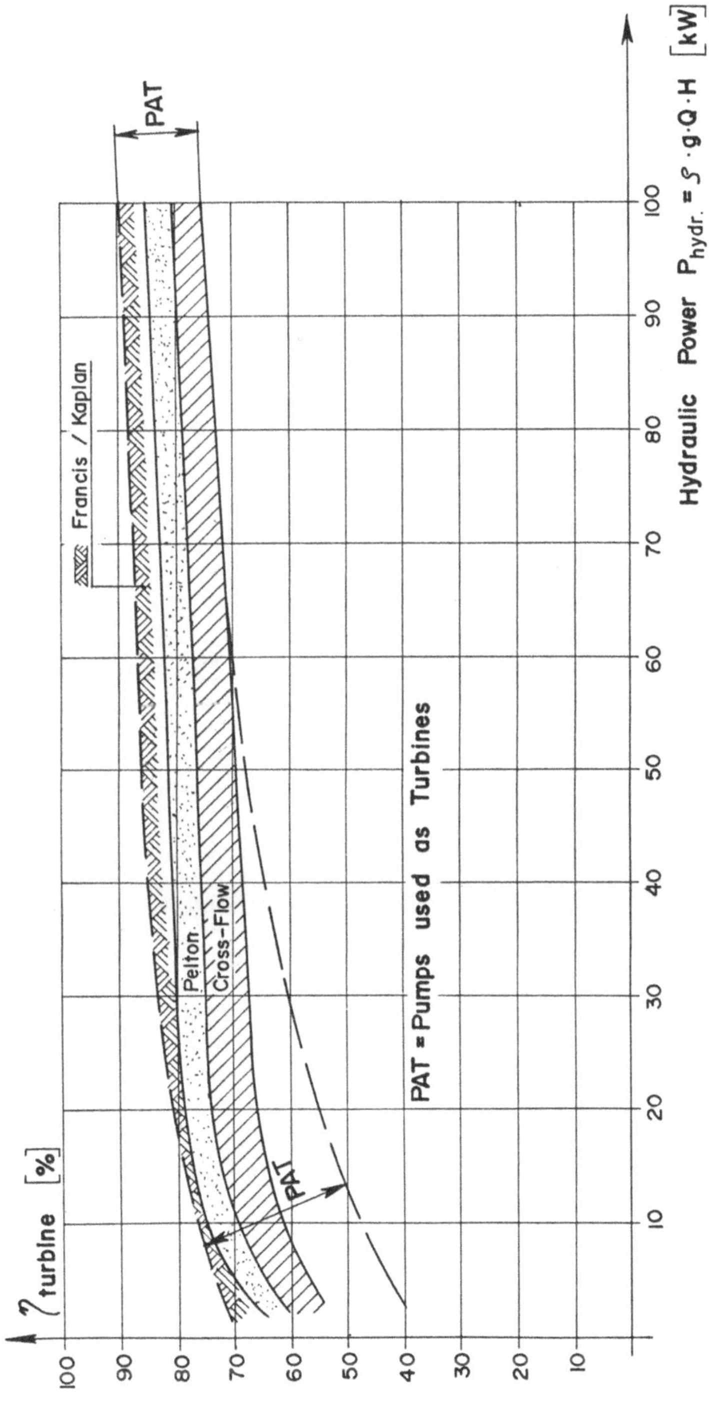

Appendix B9

Turbine Efficiencies at Part Flow

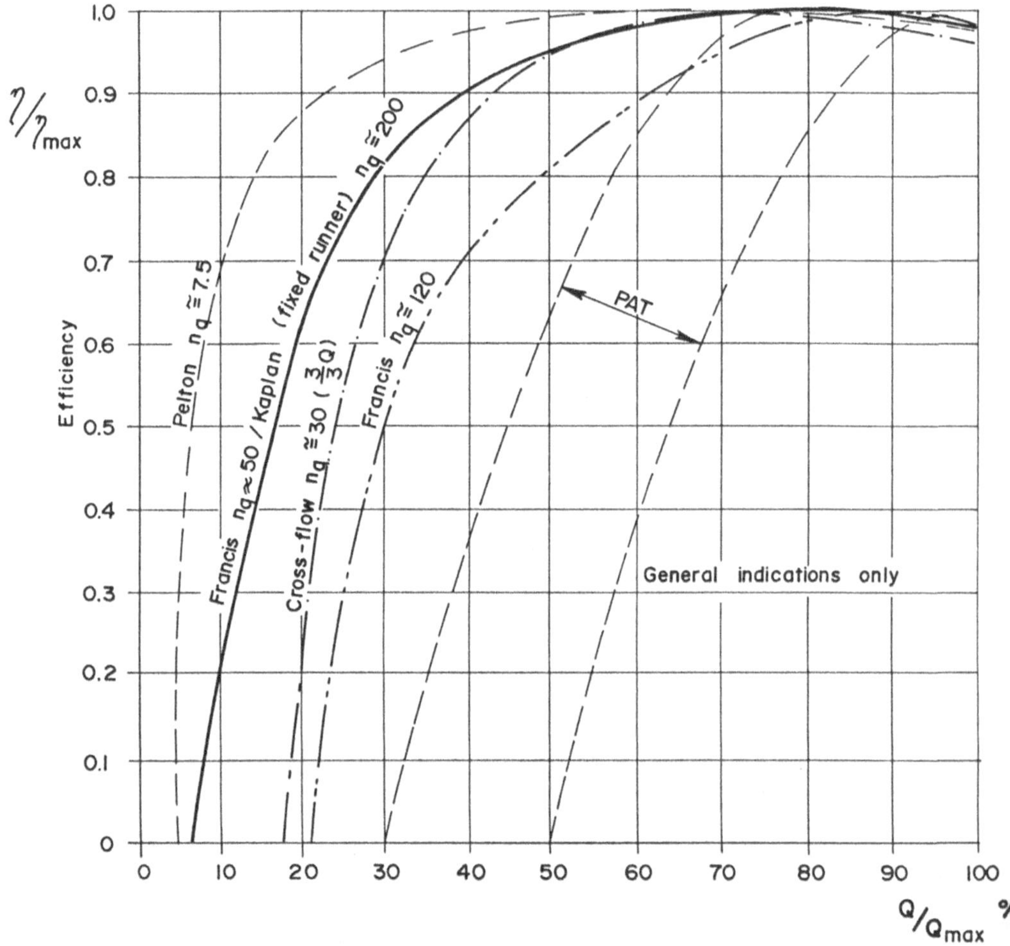

APPENDIX C

NO-LOAD AND BLOCKED-ROTOR TESTS AND DETERMINATION OF THE CIRCLE DIAGRAM

1. General

In order to determine the circle diagram of an induction machine, the following tests have to be performed:

- measurement of the resistance of the stator winding per phase;

- no-load test

- blocked-rotor test

2. Measurement of stator resistance

The resistance of the stator windings can be measured with one of the following methods:

- measuring DC voltage and DC current and calculating the resistance: $R = U / I$. In order to avoid a temperature rise of the windings during the measurement, the value of the applied current must not exceed 10 % of the nominal current.

- using an ohm-meter

The resistance (R_{T_0}) measured at ambient temperature (T_0) should be converted into the resistance at the normal operating temperature T; for copper windings:

$$\boxed{R_T = R_{T_0}\,\frac{235 + T}{235 + T_0}} \quad \text{where T and } T_0 \text{ in } {}^\circ\text{C} \tag{C1}$$

It must be assured that the resistance of the individual windings are measured; i.e., all six coil ends must be free of contact plates in the terminal box.

For further computations, the mean value of the three phase resistances will be used.

3. No-Load Test

For a squirrel-cage induction motor the test must be performed with rotating rotor while for a wound-rotor machine it must be done at standstill and in rotation. We will treat only the squirrel-cage machine since the wound-rotor motor is rarely used as a generator.

The motor is run with the load uncoupled and rated voltage supplied to the stator . Measure:

- line-to-line voltage U_{s0}

- line current I_{s0}

- three-phase power input P_0

The equipment required for the no-load test are given Figure C1. Note, the current transformers (CT) connected to the measuring instruments are only required for larger motors where currents exceed the permissible current of the ammeters.

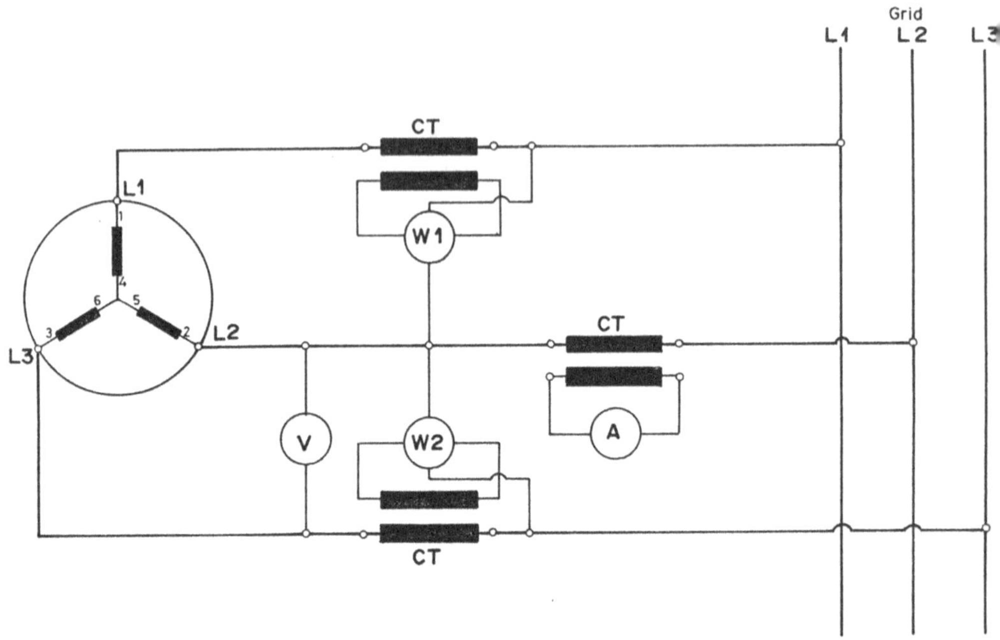

FIGURE C1 :
Arrangement for no-load test using the two-wattmeter method (see also Appendix D) and current transformers (CT)

These measurements should be repeated at varying voltage (various points between 0.2 U_n and 1.2 U_n). A varying voltage source may be composed of an adjustable transformer or, more conveniently for non-laboratory applications, of equal resistances on each line.
Varying voltage allows the no-load characteristics U_{s0} versus I_{s0} to be determined which is a measure of the saturation level of the machine (see Figure 4.3 b) on page 47). Additionally, the friction and windage losses of the motor P_{f+w} can be determined (Fig. C2). These mechanical losses cannot be considered in the circle diagram and must therefore be deducted from the input power.
Since the motor at no-load runs at very low value of slip, the induced current is low and the no-load rotor copper loss is negligible. Therefore, the input power consists of core loss P_{fe}, stator copper loss P_{Cu_s} (due to the current I_{s0}) and the friction and windage loss P_{f+w}.

$$P_0 = P_{fe} + P_{Cu_s} + P_{f+w}$$

Stator copper losses (ohmic) can be determined from the measured voltage and current values:

$$\boxed{P_{Cu_s} = 3\, I_{s0}^{\ 2} R_T} \quad \text{(Watt)} \qquad \text{(C2)}$$

Note: the resistance R of the single winding must be corrected from the temperature at the resistance measurement (≈ 20° C) to the winding temperature during the no-load test (T ≈ 35° C) using formula C1 above.
The sum (designated P_{00}) of the friction and windage losses plus the core losses can be evaluated:

$$\boxed{P_{00} = P_{cu_s} + P_{f+w} = P_0 - 3\, I_{s0}^{\,2} R_T} \qquad (C3)$$

Plotting P_{00} versus the square of the voltage U_{s0} provides a graph (≈straight line) which permits to determine the friction and windage losses: extending the graph to zero voltage, the power read on the ordinate corresponds to the friction and windage losses P_{f+w} as at no voltage, core losses are zero.

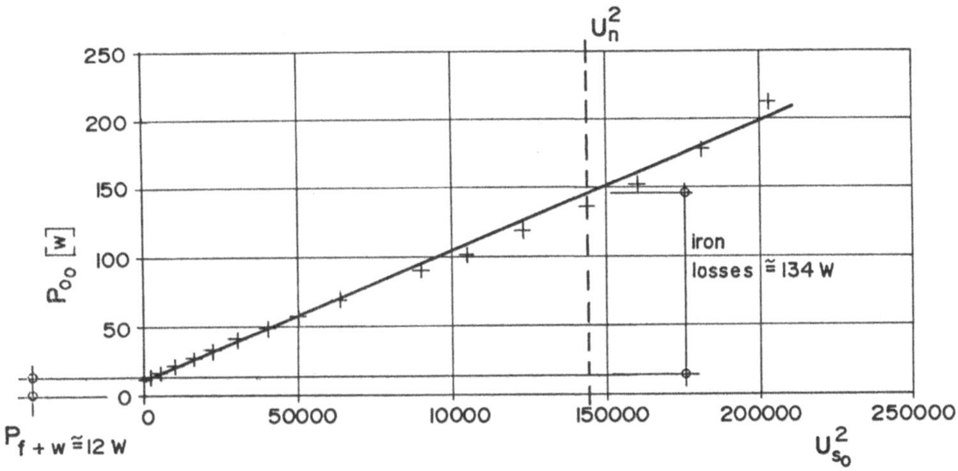

FIGURE C2 :
Determination of friction and windage losses (3-phase induction motor, 1.1 kW mec. output)

The no-load power factor to be used in the circle diagram can now be calculated (use the power, line current and line voltage values of the no-load test at rated voltage):

$$\boxed{\cos\phi_0 = \frac{P_0 - P_{f+w}}{\sqrt{3}\, U_{s0}\, I_{s0}}} \qquad (C4)$$

No-load power factors for standard induction machines are in the order of $\cos\phi_0 \approx 0.1$.

The vector with current I_{s0} at angle ϕ_0 provides the first point on the circle diagram (Point P_0 ; see Figure C4 below).

4. Blocked-Rotor Test

The blocked-rotor test is performed with the rotor blocked; i.e the rotor is prevented from rotating. This corresponds to the conditions at start-up where slip s = 100 %. Since starting currents are in the order of five to six times rated current, the test is performed at a limited voltage in order not to burn the windings during the test. Hence, the stator voltage should be reduced (by a transformer or equal resistances on each phase) to such a value that only rated stator current will flow; this will be at about 20% of rated voltage. Measure:

- line-to-line stator voltage Us_{Sc} (Sc stands for short-circuit)
- line current Is_{Sc}
- three-phase input power P_{Sc}

Since the voltage applied will not necessarily yield the exact value of rated current, the measured power and voltage values must be converted to I_n. Plot a graph with Us_{Sc} versus Is_{Sc} and another one with P_{Sc} versus the square of Is_{Sc}. Both graphs should follow a linear law through zero (may be verified if a varying voltage source is made available). Entering the graphs at rated motor current provides the values of the short-circuit voltage and power required for further computations.

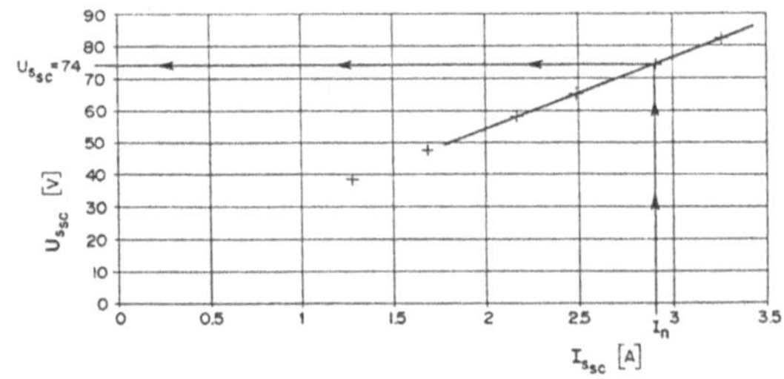

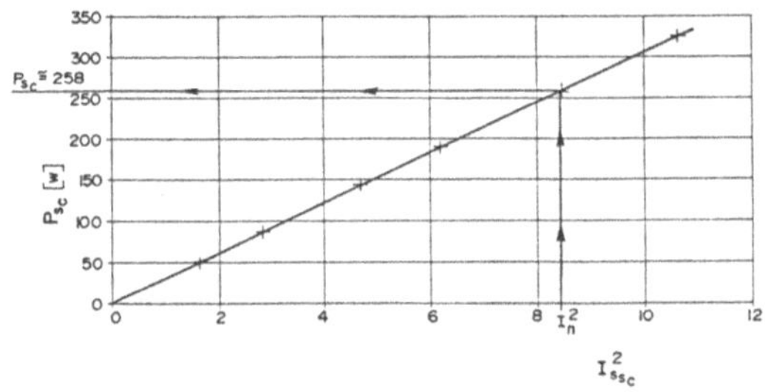

FIGURE C3 :
Blocked rotor test: determination of short-circuit (blocked-rotor) voltage and power at rated current (3-phase induction motor, 1.1 kW mec. output)

With these values (total power P_{Sc}, line current Is_{Sc} and line voltage Us_{Sc}) the short-circuit power factor $\cos\phi_{SC}$ can be calculated:

$$\boxed{\cos\phi_{Sc} = \frac{P_{Sc}}{\sqrt{3}\ Us_{Sc}\ Is_{Sc}}} \qquad (C5)$$

The blocked-rotor power factor for standard induction motors is in the order of $\cos\phi_{Sc} = 0.4 - 0.8$ (the higher values apply to small machines).
In order to plot the point P_k (corresponding to a slip value of 100 % = conditions at start up), the measured current Is_{Sc} must be extrapolated to nominal voltage:

$$\boxed{I_{st} = Is_{Sc}\left(\frac{U_n}{Us_{Sc}}\right)} \qquad (C6)$$

The point P_k of the circle diagram can now be plotted using the vector with I_{st} and angle ϕ_{Sc} (see Figure C4 below).

5. Determination of the Circle

The no-load point P_0 must be corrected to the point P_0' by taking the no-load copper losses into account. The point P_0' can be found at a distance of $I_{0\rho}$ from the abscissa on the circle (estimate the course of the circle by departing from P_0).

$$I_{0\rho} = \frac{I_0^{\,2} R_T}{U_{ph}} \qquad \text{(C7)}$$

where U_{ph} = phase voltage; $U_{ph} = \frac{U_n}{\sqrt{3}}$ (for star connection); $U_{ph} = U_{line}$ (for Δ - connection)

The value of R per phase has been measured previously (see Section 2 above). It must be corrected to the actual operating temperature T of the windings at full output ($T \approx 80 - 100^\circ$ C); use Formula C1 above).

The centre of the circle M is determined by the median perpendicular through the two points P_0' and P_k and the straight line through P_0' at an angle 2α to the abscissa:

$$\tan 2\alpha = \frac{2\, I_0 R_T}{U_{ph}} \qquad \text{(C8)}$$

Plotting the vector for rated current at full-load power factor (either measured or from rating plate data) provides a third point which should theoretically be on the circle, too. This will not always be the case. If the full-load power factor is based on measurements, i.e., more accurate than from the rating plate, the circle can be improved by determining its centre using the two median perpendiculars through points P'_0, P_k and P_n.

The performance of the machine in generator mode can be found on the lower half of the circle. Generally, the output power at nominal voltage is in most cases too optimistic since the circle diagram does not take into account the saturation effects of the magnetic circuits.

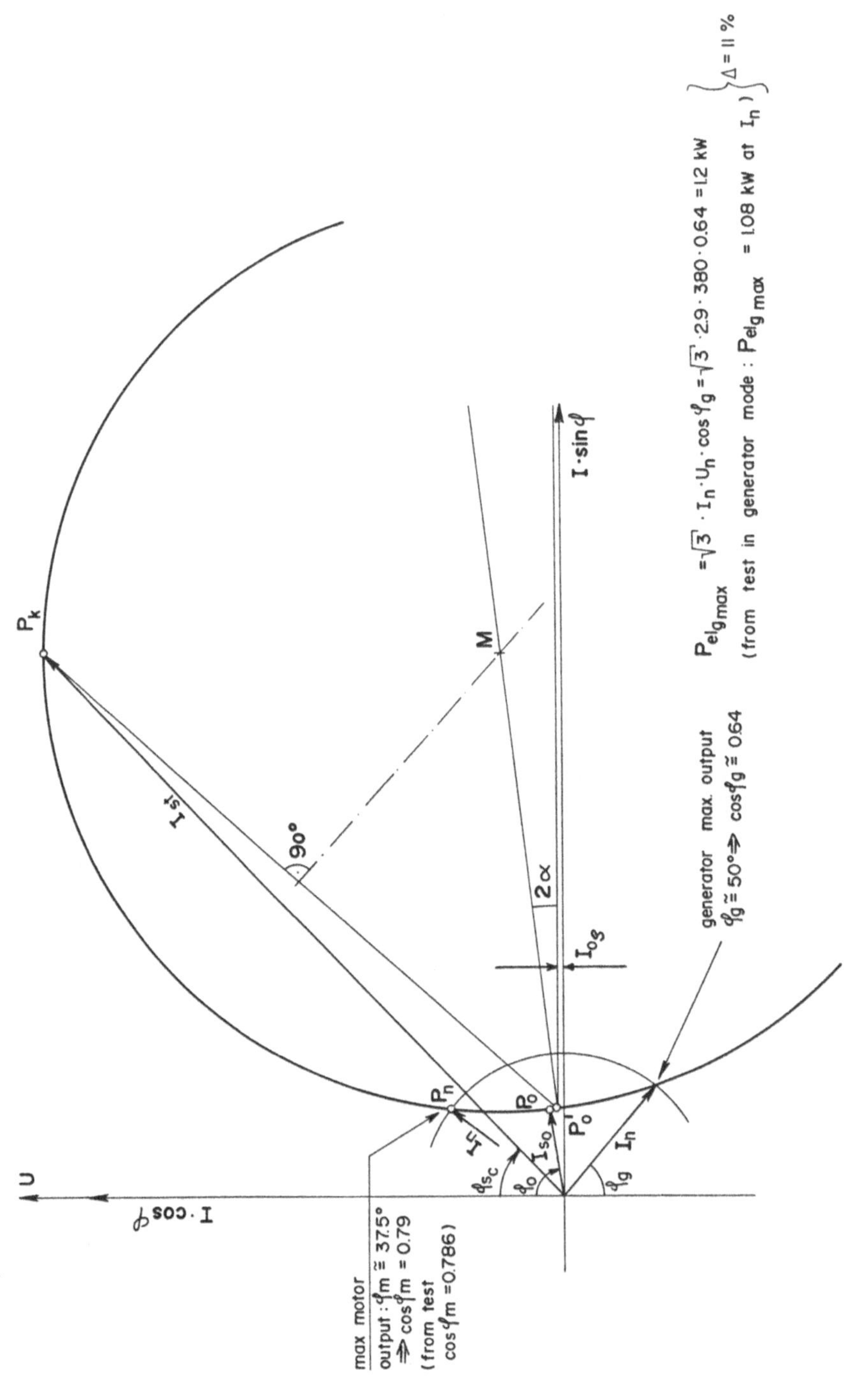

FIGURE C4 :
Circle diagram of an induction machine plotted from no-load and blocked-rotor test data (3-phase induction motor of 1.1 kW mec. output)

6. Example

6.1 Motor data (from rating plate):

- Squirrel-cage induction motor , three-phase, enclosure IP54
- Rated output : P_n = 1.1 kW at 220 / 380 V (Δ / Y), 50 Hz
- Rated Current : I_n = 5 / 2.9 A
- Rotational speed: n = 1430 rpm
- Starting current: $I_{st} / I_n = 6.1$

6.2 Motor resistances (using an Ω-meter)

Winding temperature $T_o \approx$ ambient temperature = 20° C

Terminal	Resistance [Ω]
1 / 4	5.64
2 / 5	5.68
3 / 6	5.62

mean value: $R_{T_o} = 5.65\ \Omega$

6.3 No-load test

Motor Y-connected; supply voltage 380 V (50 Hz); nominal current I_n = 2.9 A.

Measurements:

- P_o = three-phase input power
- U_{s_o} = line voltage
- I_{s_o} = line current = phase current
- T_{test} = 35° C
- $R_T = R_{T_o} \dfrac{235 + 35}{235 + 20} = 5.98\ \Omega$

No.	Us_o [V]	Is_o [A]	P_o [W]
1	51.3	0.33	15.3
2	75.6	0.39	19.0
3	101.9	0.47	25.5
4	128.1	0.57	33.3
5	150.9	0.65	40.9
6	175.2	0.75	50.9
7	202.2	0.85	61.7
8	225.1	0.94	73.0
9	253.3	1.06	87.7
10	300.7	1.28	119
11	325.0	1.41	137
12	351.5	1.57	162
13	380.0	1.79	193
14	400.8	2.00	224
15	425.7	2.31	273
16	450.8	2.72	344

$P_{Cu_s} = 3\, I_{s_o}^2\, R_T$	$U_{s_o}^2$
2.0	2632
2.7	5715
4.0	10'384
5.8	16'410
7.6	22'771
10.1	30'695
13.0	40'885
15.9	50'670
20.2	64'161
29.4	90'420
35.7	105'625
44.2	123'552
57.5	144'400
71.8	160'640
95.8	181'220
132.8	203'221

$P_{oo} = P_o - P_{Cu_s}$. The plot of P_{oo} versus $U_{s_o}^2$ is shown on Figure C2 above. We obtain: $P_{f+w} \approx 12$ W.

At rated voltage (measurement No. 13) :

$P_o - P_{f+w} = 193 - 12\ W = 181\ W$

Apparent power $S = \sqrt{3}\ I_{s_o}\ U_{s_o} = \sqrt{3} * 1.79 * 380 = 1178\ VA$

$$\cos \phi_o = \frac{P_o - P_{f+w}}{S} = \frac{181}{1178} = 0.154$$

$I_{s0}\ \cos\phi_o = 0.275\ A$

$I_{s0}\ \sin\phi_o = 1.769\ A$

6.4 Blocked-rotor test

Motor connected as for no-load test

Measurements: line voltage and line current

No.	Us_{Sc} [V]	Is_{Sc} [A]	P_{Sc} [W]
1	82.2	3.26	325
2	73.7	2.91	255
3	65.4	2.49	192
4	58.4	2.17	146
5	47.5	1.69	88
6	38.3	1.28	51

The graphs of Us_{Sc} versus Is_{Sc} and P_{Sc} versus $Is_{Sc}{}^2$ are presented in Figure C3 above.

$$\cos\phi_{Sc} = \frac{P_{Sc}}{\sqrt{3}\ Us_{Sc}\ Is_{Sc}} = \frac{258}{\sqrt{3} * 74.0 * 2.9} = 0.694$$

$I_{st} = I_{Sc}\ (U_n\ /\ Us_{Sc}) = 2.9\ (380\ /\ 74.0) = 14.9\ A\ (\hat{=}\ 5.10\ I_n)$

$I_{st}\ \cos\phi_{Sc} = 10.34\ A$

$I_{st}\ \sin\phi_{Sc} = 10.73\ A$

6.5 Determination of the circle

Stator resistance as during test (T = 35° C)

$R_T = 5.98\ \Omega$

$$I_{o\rho} = \frac{I_o^2\ R_T}{U_{ph}} = \frac{1.79^2 * 5.98}{220} = 0.087\ A$$

$$\tan2\alpha = \frac{2\ I_0\ R_T}{U_{ph}} = \frac{2 * 1.79 * 5.98}{220} = 0.097$$

$2\alpha = \arctan(0.097) = 5.56\ ^\circ$

The construction of the circle is presented in Figure C4 above.

APPENDIX D

MEASURING TECHNIQUES AND REQUIRED EQUIPMENT

1. Accuracy of Measurements

Every instrument has an inevitable indication error which is caused by imperfect manufacture (e.g., friction in bearings) or imperfect quality of materials (e.g., permanent magnets with slightly varying fields) . The error of measurements is indicated on every instrument as a **percentage of the full-scale reading**. An instrument with an accuracy of 1 % and a full-scale reading of say 500 V will therefore have an error on all measurements of 5 V. If you measure voltage of 220 V with this instrument, the measuring error will be : 5 V/220 V * 100% = 2.3 % (and not 1%). It is therefore not only important to use accurate instruments but also to select the appropriate measuring range for each measurement (always between 2/3 and full-scale reading).
Common classes of measuring instruments are the following:

Class	*admissible error of indication in ± % of full-scale value*	*Remarks*
0.1	0.1	precision measuring instruments (very expensive)
0.2	0.2	ditto class 0.1
0.5	0.5	
1	1.0	
1.5	1.5	for boards and panels
2.5	2.5	
5	5.0	

2. Voltage Measurement

A voltmeter measures the potential difference between two points of an electric circuit. It is connected in parallel to the two points. The smallest possible current should flow through the instrument in order that the measurement does not affect the normal operation of the circuit. Voltmeters therefore have a coil with very many turns of thin wire (high impedance). The higher the voltage to be measured, the higher the current pressed through the instrument. If the voltage to be measured exceeds the range of the instrument, a voltage transformer may be used.

3. Current Measurement

An ammeter measures the current flowing through a circuit; it is connected in such a way that the current must flow through the meter. Ammeters therefore must have a coil with only a few turns of thick wire. They should represent the smallest possible impedance.

The ammeter must never be connected like a voltmeter, i.e. in parallel to the source. This would create a short-circuit and invariably destroy the instrument.

An elegant way to measure currents (especially in power systems) is the use of clamp current meters. These need not be connected into the circuit. The clamp meter works on the principle that any conductor with a current flowing through it develops a magnetic field around itself which will induce voltage into the clamp. Some designs of clamp meters may also measure voltage and power factor.

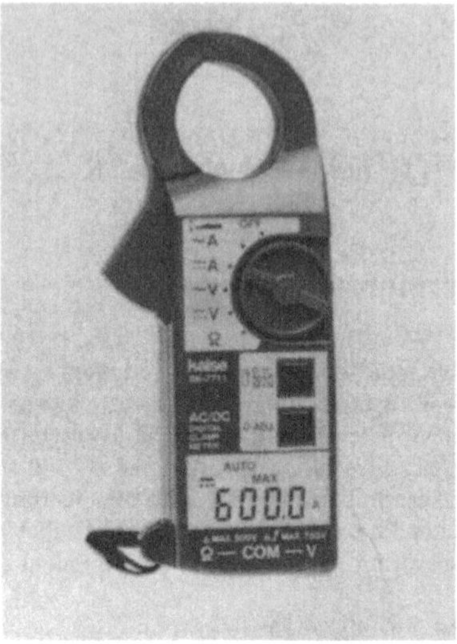

FIGURE D1 :
Clamp current meter

Built-in instruments in power stations may have to be equipped with current transformers (CT) if currents of a certain intensity have to be handled (I > 30 - 50 A). The connection of CTs is shown in Section 4.2 below.

4. Active Power Measurement by Means of Wattmeters

4.1 Principle of a Wattmeter

A wattmeter measures the power being consumed by a circuit. Current and voltage are measured simultaneously.
An analogue wattmeter comprises two coils:
- the first one (index 1) is connected in series into the circuit; the current flowing through it creates a magnetic field of induction B; the coil is connected like an ammeter;

- the second coil (index 2) is connected in parallel to the consumer (similar to a voltmeter). It is mounted in such a way that it can rotate freely with the pointer of the instrument.

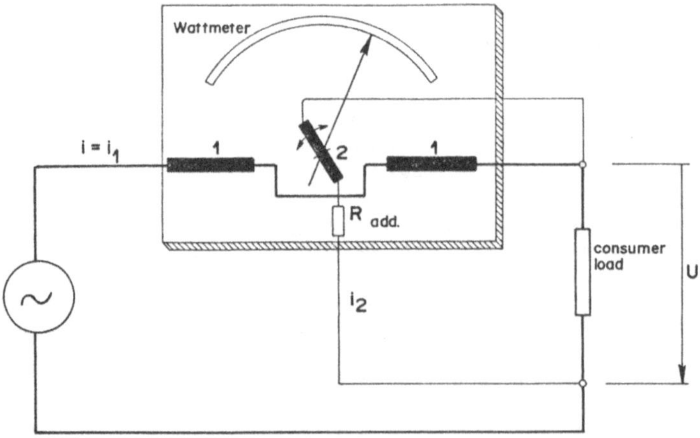

FIGURE D2 :
Principle of a Wattmeter (analogue instrument)

The interaction between the induction B created by the current flowing through the fixed coil 1 and the current i2 flowing through the movable coil 2 produces a torque which is proportional to the active power consumed by the appliance.

4.2 Current Transformer

If a wattmeter (or an ammeter) has to handle currents exceeding the maximum range of the instrument, a current transformer (CT) may be used. Figure D3 below shows the connection of the wattmeter, the ammeter and the transformer.

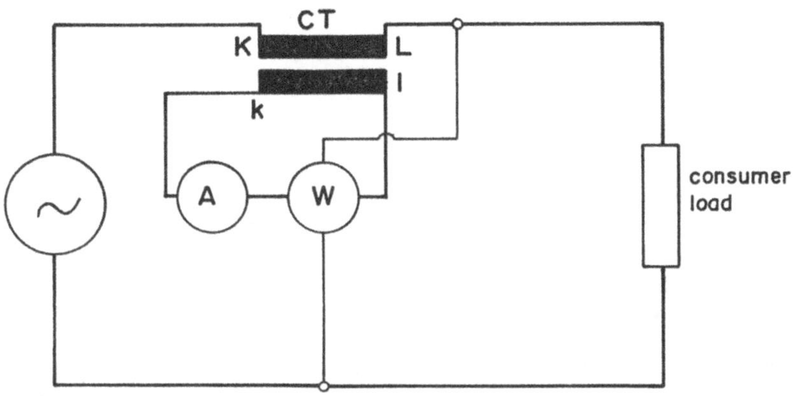

FIGURE D3 :
Connection of a wattmeter and an ammeter in conjunction with a current transformer (CT)

4.3 The Constant of a Wattmeter

An instrument may have different terminals with different current and voltage ranges but only a single scale on the (analogue) display. The reading on the scale must therefore be multiplied by a constant c_w in order to take the selected (voltage and current) range into account.

$P = \alpha\, c_w$ where α = reading in divisions and c_w = constant (Watt/division)

$$c_w = \frac{\text{range U} * \text{range I}}{\text{total number of divisions}} \qquad \text{(D1)}$$

Figure D4 below shows how it works.

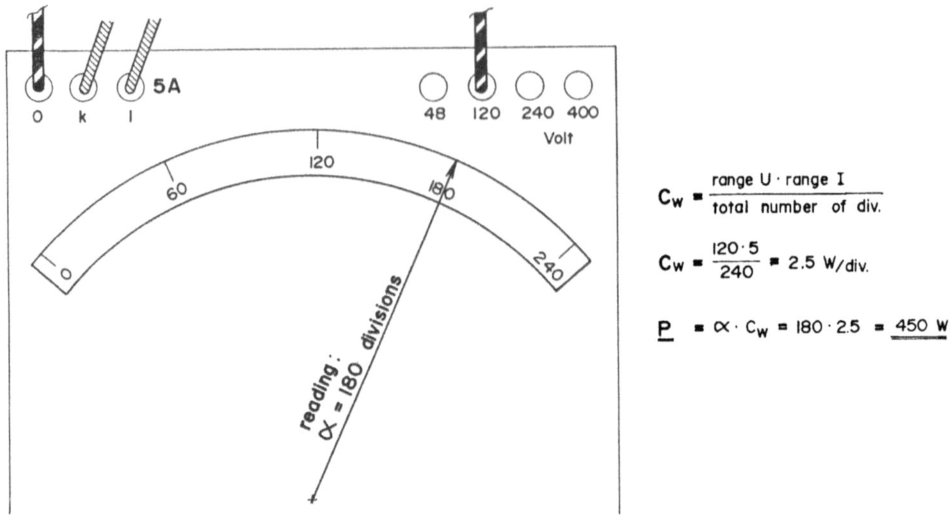

FIGURE D4 :
Example of using the wattmeter constant

If a current transformer is used, the wattmeter constant must be multiplied by the transformer ratio c_i (e.g., 100 / 5 A) to obtain active power:

$$P = c_w \; c_i \; \alpha$$

4.4 Active Power Measurement in a Three-Phase System

A three phase system may be considered as three single-phase systems working on common conductors. The total power of a three-phase system is therefore the algebraic sum of the power consumed (or furnished) by each phase:

$$P_{tot} = \pm P_1 \; \pm P_2 \; \pm P_3$$

4.4.1 By Means of a Single Wattmeter

This method is only valid if the **system is perfectly balanced**. Voltage is measured between phase and neutral conductors (see Figure D5 a). In a three-wire system without neutral, an artificial neutral must be created by means of three resistances, the third one being the resistance R_w of the wattmeter. It must be ensured that $R_1 = R_2 = R_w$ (see Figure D5b)

Total active power of the balanced three-phase system then becomes:

$$\boxed{P_{tot} = 3\,P_w} \qquad \text{(D2)}$$

where P_w is the power measured on the single wattmeter.

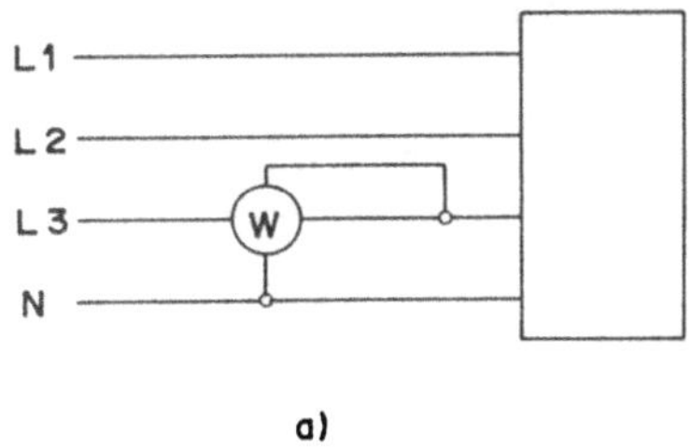

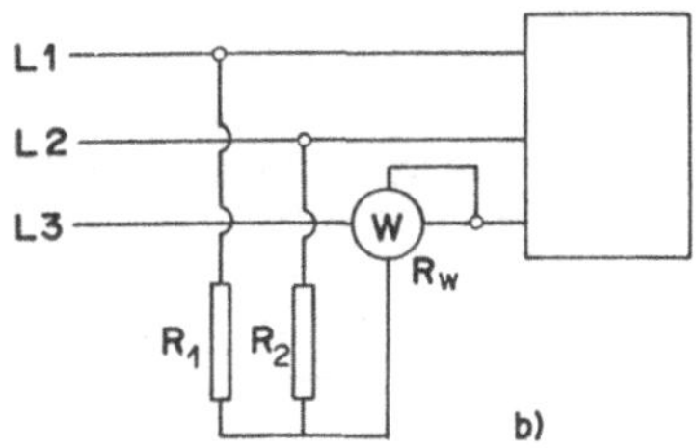

FIGURE D5 :
Measurement of three-phase active power by means of a single wattmeter
a) with neutral b) without neutral

4.4.2 By Means of Three Wattmeters

This method may be used if the system is imbalanced. If the neutral is not available, voltage measurements could be done as indicated in Figure D6. In this case, the measurement is only correct if the three wattmeters have identical characteristics (especially as far as the resistance of the voltage coils are concerned). If therefore $cw_1 = cw_2 = cw_3 = cw$, then

$$P_{tot} = \pm P_1 \pm P_2 \pm P_3$$

$$= \pm cw_1 \alpha_1 \pm cw_2 \alpha_2 \pm cw_3 \alpha_3$$

$$\boxed{P_{tot} = c_w (\pm\alpha_1 \pm\alpha_2 \pm\alpha_3)} \qquad \text{(D3)}$$

It is important that the sense of rotation of the wattmeter pointer is verified. If it moves anti-clockwise (towards negative values) the current terminals must be inverted to obtain a reading; this means that power on that phase is negative and α in Formula D3 will become negative, too. Depending on the power factor, one or two of the three wattmeters could indicate negative values while total power would still remain positive.

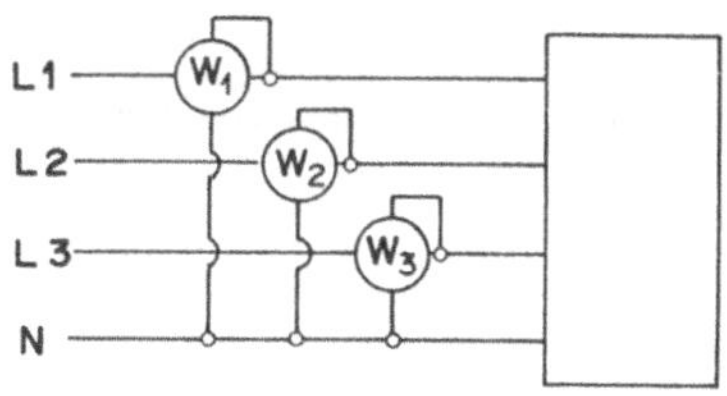

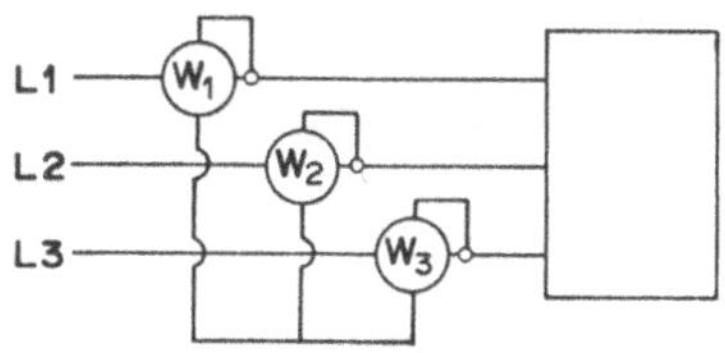

FIGURE D6 :
Measurement of three-phase power by means of three wattmeters a) with neutral b) without neutral

4.4.3 By Means of Two Wattmeters - Connection according to Aron

This method may be used for imbalanced systems as well as for systems with and without neutral. **In any case, the neutral must be disconnected from the load whose power consumption is to be measured.** The total power is the algebraic sum of the individual power indications of the two instruments:

$$P_{tot} = \pm P_1 \pm P_2 \qquad \text{(D4)}$$

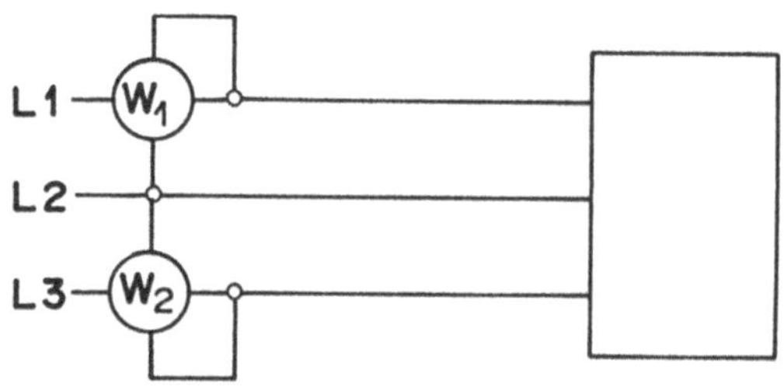

FIGURE D7 :
Measurement of three-phase power by means of two wattmeters (according to Aron)

5. Active Power Measurement by Means of a kW-Hour Meter

Generators of MHPs operating in parallel to a large grid are usually equipped with a kW-hour counter measuring active power delivered to the grid. Using a simple stop-watch, the active power generated can be determined:

- determine the meter constant X (= number of revolutions of the disc per kWh)

- measure the time T in seconds for N revolutions of the disc (T should be about one or two minutes)

- calculate active power delivered:

$$P = \frac{3600\ N}{X\ T} \qquad \text{(D5)}$$

6. Portable Instruments - Multimeters

Digital measuring instruments have become increasingly attractive not only because they combine practically all measuring tasks in one handy instrument but also because of their high accuracy at a fairly low price. For testing and commissioning IMAGs of the small power range, multimeters with current, voltage and frequency indications can be a very helpful tool. Some instruments even provide for capacitance and resistances to be checked.

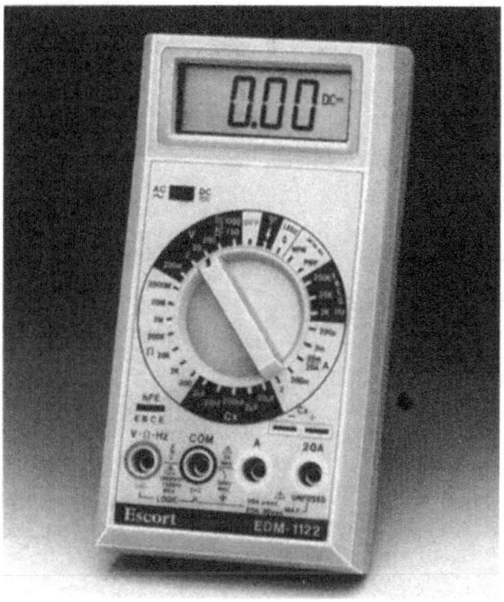

FIGURE D8 :
Multimeter

APPENDIX E

LIST OF MANUFACTURERS OF INDUCTION GENERATOR CONTROLLERS (IGC)

- ITDG
Myson House
Railway Terrace
Rugby CV 21 3 HT United Kingdom Phone 0788 / 60631
Fax 0788 / 540270

- Mr. Nigel Smith
Dept. of Electrical Engineering
Trent Polytechnic
Burton Street
Nottingham NG1 4BU United Kingdom Phone 0602 / 418 418
Fax 0602 / 484 266

- Ecole Polytechnique Fédérale, Lausanne (EPFL)
Mr. Jacques Dos Ghali
Electricity Dept.
ELG - Ecublens
1015 Lausanne Switzerland Phone +21 / 693 26 36
Fax +21 / 693 26 87

- THES Small Hydro Controls (Thomson and Howe Energy Systems Inc.)
Site 17, Box 2, S.S.1
Kimberley, British Columbia
V1A 2Y3 Canada Phone (604) 427 - 4326
Fax (604) 427 - 7723 or (604) 427 - 4326

- E C S (Energy Control Systems Limited)
3 Cranford Hall, Stillorgan Road
Donnybrook
Dublin 4 Ireland Phone + 353 .1. 283 9182
Fax + 353. 1. 780 282 or + 353. 1. 283 9182

APPENDIX F

REFERENCES

Books

Matsch, Leander W.
Morgan, J. Derald
Electromagnetic and Electromechanical Machines
Third Edition John Wiley & Sons, Inc., 1987

Fischer, Rolf
Elektrische Maschinen
ISBN 3-446-14562-1 Carl Hanser Verlag, München, 6. Auflage 1986

Boy/Flachmann/Mai
Elektrische Maschinen und Steuerungstechnik
ISBN 3-8023-0725-9 Vogel Verlag, Würzburg, 8. Auflage 1990

Ryff, Peter F.
Electric Machinery
ISBN 0-13-248691-1 Prentice Hall International Editions, London 1988

Articles

Harz, Hermann
Kondensatorerregte Asynchrongeneratoren
Siemens-Zeitschrift 25, Heft 4, S. 199 Sept. 1951

Schroeder, Dr.-Ing. J.W.
Der kondensatorerregte Asynchrongenerator
Technische Rundschau Nr. 54 vom 27. 12. 1963

Pereira, L.
Induction Generators for Small Hydro Plants
Water Power & Dam Construction Nov. 1981

Bhattacharya, J.L.
Woodward, Prof. J.L.
Excitation Balancing of a Self-Excited Induction Generator for Maximum Power Output
IEE Proceedings, Vol. 135,Pt. C, No. 2, March 1988

Germond, A.
Morf, J.-J.
Dangers des installations électriques
Résponsabilités et mesures de protection selon les règles suisses
L'Indicateur Industriel - revue de la technique européenne, nos. 9 et 10/1973

Smith, N.
A low-cost electronic load controller for stand alone water-powered induction generators
GTZ: Simple governing systems for micro-hydro, Conference, Frankfurt, 1988

J. Dos Ghali
L. Cardoletti
J.P. Ludwig
Charge adaptative
Internal report, EPF Lausanne, LEME 1990

GPSR Compliance

The European Union's (EU) General Product Safety Regulation (GPSR) is a set of rules that requires consumer products to be safe and our obligations to ensure this.

If you have any concerns about our products, you can contact us on ProductSafety@springernature.com

In case Publisher is established outside the EU, the EU authorized representative is:

Springer Nature Customer Service Center GmbH
Europaplatz 3
69115 Heidelberg, Germany

Batch number: 10577876

Printed by Printforce, the Netherlands